Ben Stacy Jerrik (Ed.)

Wasp-15B

Ben Stacy Jerrik (Ed.)

Wasp-15B

Subgiant, Star, Stellar Classification, Constellation

Part Press

Imprint

Publisher:
Part Press is a trademark of
International Book Market Service Ltd., 17 Rue Meldrum, Beau Bassin, 1713-01 Mauritius
Email: info@bookmarketservice.com
Website: www.bookmarketservice.com

Published in 2013

Printed in: U.S.A., U.K., Germany. This book was not produced in Mauritius.

ISBN: 978-613-9-12758-0

Contents

Articles

References

Viscount_Hanworth

Viscount Hanworth, of Hanworth in the County of Middlesex, is a title in the Peerage of the United Kingdom. It was created on 17 January 1936 for the lawyer, Conservative Member of Parliament and former Master of the Rolls, Ernest Pollock, 1st Baron Hanworth. He had already been created a **Baronet**, of Hanworth in the County of Middlesex, on 27 November 1922, in the Baronetage of the United Kingdom, and **Baron Hanworth**, of Hanworth in the County of Middlesex, on 28 January 1926, in the Peerage of the United Kingdom. Pollock was the fifth son of George Pollock, fourth son of Sir Frederick Pollock, 1st Baronet, of Hatton (elder brother of Field Marshal Sir George Pollock, 1st Baronet, of The Khyber Pass). As of 2012 the titles are held by the first Viscount's great-grandson, the third Viscount, who succeeded his father in 1996. As a descendant of Sir Frederick Pollock, 1st Baronet, of Hatton, he is also in remainder to this title, which is currently held by his kinsman Sir George Frederick Pollock, 5th Baronet.

Ernest Pollock,
1st Viscount Hanworth.

Viscounts Hanworth (1936)

- Ernest Murray Pollock, 1st Viscount Hanworth (1861–1936)
    - Charles Thomas Anderson Pollock (d. 1918)
- David Bertram Pollock, 2nd Viscount Hanworth (1916–1996)
- David Stephen Geoffrey Pollock, 3rd Viscount Hanworth (b. 1946)

The heir presumptive is the present holder's nephew Harold William Charles Pollock (b. 1988)

See also

- Pollock Baronets, of Hatton, and Pollock Baronets, of The Khyber Pass

References

- Kidd, Charles, Williamson, David (editors). *Debrett's Peerage and Baronetage* (1990 edition). New York: St Martin's Press, 1990.
- Leigh Rayment's Peerage Pages [1]

External links

- Hansard 1803–2005: contributions in Parliament by David Bertram Pollock, 2nd Viscount Hanworth [2]

References

[1] http://www.leighrayment.com/
[2] http://hansard.millbanksystems.com/people/mr-david-pollock-1

Tony_Benn

The Right Honourable **Tony Benn**		
[[file:Tony Benn2.jpg		alt=]]
President of the Stop the War Coalition		
Incumbent		
Assumed office 21 September 2001		
Vice President	Lindsey German	
Preceded by	Office created	
Secretary of State for Energy		
In office 10 June 1975 – 4 May 1979		
Prime Minister	Harold Wilson James Callaghan	
Preceded by	Eric Varley	
Succeeded by	David Howell	
Secretary of State for Industry		
In office 5 March 1974 – 10 June 1975		
Prime Minister	Harold Wilson	
Preceded by	Peter Walker (at DTI)	
Succeeded by	Eric Varley	
Chairman of the Labour Party		
In office 20 September 1971 – 25 September 1972		
Leader	Harold Wilson	
Preceded by	Ian Mikardo	
Succeeded by	William Simpson	
Minister of Technology		
In office 4 July 1966 – 19 June 1970		
Prime Minister	Harold Wilson	
Preceded by	Frank Cousins	
Succeeded by	Geoffrey Rippon	
Postmaster General		

In office 15 October 1964 – 4 July 1966		
Prime Minister	Harold Wilson	
Preceded by	Reginald Bevins	
Succeeded by	Edward Short	
Member of Parliament **for Chesterfield**		
In office 1 March 1984 – 7 June 2001		
Preceded by	Eric Varley	
Succeeded by	Paul Holmes	
Majority	24,633 (46.5%)	
Member of Parliament **for Bristol South East**		
In office 20 August 1963 – 9 June 1983		
Preceded by	Malcolm St Clair	
Succeeded by	Constituency Abolished	
Majority	1,890 (3.5%)	
In office 30 November 1950 – 17 November 1960		
Preceded by	Stafford Cripps	
Succeeded by	Malcolm St Clair	
Majority	13,044 (39%)	
Personal details		
Born	Anthony Neil Wedgwood Benn3 April 1925Marylebone, London, United Kingdom	
Nationality	British	
Political party	Labour	
Spouse(s)	Caroline DeCamp (m. 1949–2000, her death)	
Children	Stephen, Hilary, Melissa, Joshua	
Alma mater	New College, Oxford	
Religion	Agnostic[1]	
Website	Official website [2]	
Military service		
Service/branch	Royal Air Force	
Rank	Pilot Officer	
Battles/wars	World War II	

Anthony Neil Wedgwood Benn PC (born 3 April 1925), formerly **2nd Viscount Stansgate**, is a retired British Labour Party politician who was a Member of Parliament (MP) for 50 years and a Cabinet Minister under Harold Wilson and James Callaghan.

Benn's campaign to renounce his hereditary peerage was instrumental in the creation of the Peerage Act 1963.[3] In the Labour Government of 1964–1970 he served first as Postmaster General, where he oversaw the opening of the Post Office Tower, and later as a notably "technocratic" Minister of Technology. In the period when the Labour Party was in Opposition, for a year he was the Chairman of the Labour Party. In the Labour Government of 1974–1979 he returned to the Cabinet, initially as Secretary of State for Industry, before being made Secretary of State for Energy, retaining his post when James Callaghan replaced Wilson as Prime Minister. During the Labour Party's time in Opposition during the 1980s, he was seen as the party's prominent figure on the left, and the term "Bennite" has come to be used in Britain for someone of a more radical left-wing position.[4]

Benn has topped several polls as the most popular politician in Britain.[5] He has been described as "one of the few UK politicians to have become more left-wing after holding ministerial office."[6] Since leaving Parliament, Benn has become involved in the grass-roots politics of demonstrations and meetings, and has been the President of the Stop the War Coalition for the last decade. He has been a vegetarian since the 1970s.

Biography

Early life and family

Benn was born in London on 3 April 1925.[7] Benn's paternal grandfather was John Benn, a successful politician who was created a baronet in 1914, and his father William Wedgwood Benn was a Liberal Member of Parliament who later crossed the floor to the Labour Party. He was appointed Secretary of State for India by Ramsay MacDonald in 1929, a position he held until 1931. He was elevated to the House of Lords with the title of Viscount Stansgate in 1941; the new wartime coalition government was short of working Labour peers in the upper house.[8] From 1945 to 1946, he was the Secretary of State for Air in the first majority Labour Government.

Both his grandfathers, John Benn (who founded a publishing company)[9] and Daniel Holmes, were also Liberal MPs (respectively, for Tower Hamlets, Devonport and Glasgow Govan).[10] Benn's contact with leading politicians of the day dates back to his earliest years; he met Ramsay MacDonald when he was five,[11] David Lloyd George when he was 12 and Mahatma Gandhi in 1931, while his father was Secretary of State for India.

Benn's mother, Margaret Wedgwood Benn (née Holmes) (1897–1991), was a dedicated theologian, feminist and the founder President of the Congregational Federation. She was a member of the *League of the Church Militant*, which was the predecessor of the *Movement for the Ordination of Women* – in 1925 she was rebuked by Randall Thomas Davidson, the Archbishop of Canterbury, for advocating the ordination of women. His mother's theology had a profound influence on Benn, as she taught him that the stories in the Bible were based around the struggle between the prophets and the kings and that he ought in his life to support the prophets over the kings, who had power, as the prophets taught righteousness.[12]

Benn went to Westminster School and studied at New College, Oxford, where he read Philosophy, Politics and Economics and was elected President of the Oxford Union in 1947. In later life, Benn attempted to remove public references to his private education from *Who's Who*; in the 1975 edition his entry stated "Education—still in progress". In the 1976 edition, almost all details were omitted save for his name, jobs as a Member of Parliament and as a Government Minister, and address; the publishers confirmed that Benn had sent back the draft entry with everything else struck through.[13] In the 1977 edition, Benn's entry disappeared entirely.[14] In October 1973 he announced on BBC Radio that he wished to be known as Mr Tony Benn rather than as Anthony Wedgwood Benn, and his book *Speeches* from 1974 is credited to "Tony Benn".

In July 1943, Benn joined the Royal Air Force.[15] His father and brother Michael (who was later killed in an accident) were already serving in the RAF. Whilst holding the rank of pilot officer, Benn served as a pilot in South Africa and Rhodesia.[16]

Benn met Caroline Middleton DeCamp (born 13 October 1926, Cincinnati, Ohio) over tea at Worcester College in 1949 and nine days later he proposed to her on a park bench in the city. Later, he bought the bench from Oxford City

Council and installed it in the garden of their home in Holland Park. Tony and Caroline had four children – Stephen, Hilary, Melissa and Joshua, and ten grandchildren. Caroline Benn died of cancer on 22 November 2000, aged 74, after a prominent career as an educationalist.[17]

Benn's children have been active in politics; his first son Stephen was an elected Member of the Inner London Education Authority from 1986 to 1990. His second son Hilary was a councillor in London, and stood for Parliament in 1983 and 1987, becoming the Labour MP for Leeds Central in 1999. He was Secretary of State for International Development from 2003 to 2007, and then Secretary of State for Environment, Food and Rural Affairs until 2010. This makes him the third generation of his family to have been a member of the Cabinet, a rare distinction for a modern political family in Britain. Benn's granddaughter Emily Benn failed to win East Worthing and Shoreham in 2010,[18] becoming the Labour Party's youngest-ever candidate in the process.[19] Benn is a first cousin once removed of the late actress Dame Margaret Rutherford.[20]

Member of Parliament

Following the Second World War Benn worked briefly as a BBC Radio producer. On 1 November 1950, he was unexpectedly selected to succeed Sir Stafford Cripps as the Labour candidate for Bristol South East, after Cripps stood down because of ill-health. He won the seat in a by-election on 30 November 1950.[21] Anthony Crosland helped him get the seat as he was the MP for nearby South Gloucestershire at the time. Upon taking the oath on 4 December 1950[22] Benn became the youngest MP, or "Baby of the House" for one day, being succeeded by Thomas Teevan, who was two years younger but took his oath a day later.[23] He became the "Baby" again in 1951, when Teevan was not re-elected. In the 1950s, Benn held middle-of-the-road or soft left views, and was not associated with the young left wing group around Aneurin Bevan.[24]

Peerage reform

Benn's father had been created Viscount Stansgate in 1942 when Winston Churchill increased the number of Labour peers to aid political work in the House of Lords; at this time, Benn's elder brother Michael was intending to enter the priesthood and had no objections to inheriting a peerage. However, Michael was later killed in an accident while on active service in the Second World War, and this left Benn as the heir to the peerage. He made several attempts to renounce the succession, but they were unsuccessful.[24]

In November 1960 Viscount Stansgate died and Benn automatically became a peer and was thus prevented from sitting in the House of Commons. Insisting on his right to abandon his peerage, Benn fought to retain his seat in a by-election caused by his succession on 4 May 1961. Although he was disqualified from taking his seat, he was re-elected. An election court found that the voters were fully aware that Benn was disqualified, and declared the seat won by the Conservative runner-up, Malcolm St Clair, who was at the time also the heir presumptive to a peerage.[3]

Benn continued his campaign outside Parliament, and eventually the Conservative Government of the time accepted the need for a change in the law.[25] The Peerage Act 1963, allowing renunciation of peerages, became law shortly after 6 pm on 31 July 1963. Benn was the first peer to renounce his title, at 6.22 pm that day. St Clair, fulfilling a promise he had made at the time of his election, then accepted the office of Stewardship of the Manor of Northstead, disqualifying himself from the House (outright resignation not being possible). Benn returned to the Commons after winning a by-election on 20 August 1963.[24]

In government (1964–1970)

In the 1964 Government of Harold Wilson, Benn was Postmaster General, where he oversaw the opening of the Post Office Tower, then the UK's tallest building, and the creations of the Post Bus service and Girobank. He proposed issuing stamps without the Sovereign's head, but this met with private opposition from the Queen. Instead, the portrait was reduced to a small profile in silhouette, a format that is still used on commemorative stamps.[26] Benn also led the government's campaign to close down the many off-shore pirate radio stations of the time, a campaign that formed the centrepiece of the 2009 film *The Boat That Rocked*, and he was responsible for introducing the Marine Broadcasting Offences Bill.[27] By the time the bill became law in 1967, Benn had been promoted to Minister of Technology, which included responsibility for the development of Concorde and the formation of International Computers Ltd. The period also saw government involvement in industrial rationalisation, and the merger of several car companies to form British Leyland. [28] Labour lost the 1970 election to Edward Heath's Conservatives, and upon Heath's application to join the European Economic Community, Benn campaigned in favour of a referendum on the UK's membership. The Shadow Cabinet voted to support a referendum on 29 March 1972, and as a result Roy Jenkins resigned as Deputy Leader of the Labour Party.

In government (1974–1979)

In the Labour Government of 1974 Benn was Secretary of State for Industry, increased nationalised industry pay, provided better terms and conditions for workers such as the Health and Safety at Work Act 1974 and set up worker cooperatives to motivate and reform struggling industries, the best known being at Meriden, outside Birmingham which kept Triumph Motorcycles in production until 1983. In 1975 he was appointed Secretary of State for Energy, immediately following his unsuccessful campaign for a "No" vote in the referendum on the UK's continued membership of the European Community (Common Market). Later in his diary (25 October 1977) Benn wrote that he "loathed" the EEC; he claimed it was "bureaucratic and centralised" and "of course it is really dominated by Germany. All the Common Market countries except the UK have been occupied by Germany, and they have this mixed feeling of hatred and subservience towards the Germans".[29]

Harold Wilson resigned as Leader of the Labour Party and Prime Minister in March 1976. In the resulting leadership contest Benn came fourth out of the six cabinet ministers who stood — he withdrew as 11.8% of colleagues voted for him in the first ballot. Benn withdrew from the second ballot and supported Michael Foot, which James Callaghan eventually won. Despite not receiving his support in the second and third rounds of the vote, Callaghan kept Benn as Energy Secretary. In autumn 1976 there was a sterling crisis, and Chancellor of the Exchequer Denis Healey sought a loan from the International Monetary Fund. Underlining a wish to counter international market forces which seemed to penalise a larger welfare state, Benn publicly circulated the divided Cabinet minutes from the 1931 National Labour Government of Ramsay MacDonald tacking the Great Depression which cut unemployment benefits to obtain a loan from American bankers. As he highlighted, these minutes resulted in the 1931 split of the Labour Party into National Labour (Organisation) and other Labour Party members. Callaghan allowed Benn to put forward his "alternative economic strategy", which consisted of a seft sufficient economy less dependent on low-rate fresh borrowing but the AES, which according to opponents, would have led to a "siege economy", was rejected by the Cabinet.[30]

Move to the Left

By the end of the 1970s, Benn had migrated to the left wing of the Labour Party. He attributed this political shift to his experience as a Cabinet Minister in the 1964–1970 Labour Government. Benn wrote:

> As a minister, I experienced the power of industrialists and bankers to get their way by use of the crudest form of economic pressure, even blackmail, against a Labour Government. Compared to this, the pressure brought to bear in industrial disputes is minuscule. This power was revealed even more clearly in 1976 when the IMF secured cuts in our public expenditure. These lessons led me to the conclusion that the UK is only superficially governed by MPs and the voters who elect them. Parliamentary democracy is, in truth, little more than a means of securing a periodical change in the management team, which is then allowed to preside over a system that remains in essence intact. If the British people were ever to ask themselves what power they truly enjoyed under our political system they would be amazed to discover how little it is, and some new Chartist agitation might be born and might quickly gather momentum.[31]

Benn's philosophy consisted of a form of syndicalism, state planning where necessary to ensure national competitiveness, greater democracy in the structures of the Labour Party and observance of Party Conference decisions;[32] he was vilified by most of the press which supported the young Margaret Thatcher — his opponents implied and stated that a Benn-led Labour Government would implement a type of Eastern European socialism.[33] Benn was overwhelmingly popular with Labour activists: a survey of delegates at the Labour Party Conference in 1978 found that by large margins they supported Benn for the leadership and many Bennite policies.[34]

He publicly supported Sinn Féin and the unification of Ireland, although in 2005 he suggested to Sinn Féin leaders that it abandon its long-standing policy of not taking seats at Westminster. Sinn Féin argue that to do so would recognise Britain's claim over Northern Ireland, and the Sinn Féin constitution prevented its elected members from taking their seats in any British-created institution.[35]

In opposition

In a keynote speech to the Labour Party Conference of 1980, shortly before the resignation of party leader James Callaghan and election of Michael Foot as successor, Benn outlined what he envisaged the next Labour Government would do. "Within days", a Labour Government would gain powers to nationalise industries, control capital and implement industrial democracy; "within weeks", all powers from Brussels would be returned to Westminster, and abolish the House of Lords by creating one thousand peers and then abolishing the peerage. Benn received tumultuous applause.[36]

In 1981, he stood against incumbent Denis Healey for Deputy Leader of the Labour Party, disregarding the appeal from Michael Foot to either stand for the leadership or abstain from inflaming the party's divisions. Benn defended his decision with insistence that it was "not about personalities, but about policies." The contest was extremely closely fought, and Healey won by a margin of barely 1%. The decision of several moderate left-wing MPs, including Neil Kinnock, to abstain triggered the split of the Campaign Group from the Left of the Tribune Group.[37]

Benn speaking at the Glastonbury Festival in 2008

After Argentina invaded the Falkland Islands in April 1982, Benn argued that the dispute should be settled by the United Nations and that the British Government should not send a task force to recapture the islands. The task force was sent and the Falklands were soon back in British control. In a subsequent debate in the Commons, Benn's demand for "a full analysis of the costs in life, equipment and money in this tragic and unnecessary war" was rejected by Margaret Thatcher, who, apparently unaware of Benn's service in World War Two and the loss of his brother, stated that "he would not enjoy the freedom of speech that he put to such excellent use unless people had been prepared to fight for it".[38]

For the 1983 election Benn's Bristol South East constituency was abolished by boundary changes, and he lost to Michael Cocks the battle to stand in the new seat of Bristol South. Rejecting offers from the new seat of Livingston in Scotland, Benn contested Bristol East, losing to Conservative candidate Jonathan Sayeed in what was perceived to be a shock result. He was selected for the next Labour seat to fall vacant, and was elected for Chesterfield in a by-election after Eric Varley resigned to head Coalite. On the day of the by-election, 1 March 1984, *The Sun* newspaper ran a hostile feature article "Benn on the Couch", which purported to be the opinions of an American psychiatrist.[39] In the period since Benn's defeat in Bristol, Michael Foot had stepped down after the general election in June 1983 (which saw Labour return a mere 209 MPs) and was succeeded in October of that year by Neil Kinnock.[40]

Benn was a prominent supporter of the 1984–1985 UK miners' strike and of his long-standing friend, the National Union of Mineworkers leader Arthur Scargill. However, some miners considered Benn's 1977 industry reforms to have caused problems during the strike; firstly, that they led to huge wage differences and distrust between miners of different regions; and secondly that the controversy over balloting miners for these reforms made it unclear as to whether a ballot was needed for a strike or whether it could be deemed as a "regional matter" in the same way that the 1977 reforms had been.[41] [42]

In June 1985, three months after the miners admitted defeat and ended their strike, Benn introduced the Miners' Amnesty (General Pardon) Bill into the Commons, which would have extended an amnesty to all miners imprisoned during the strike. This would have included two men convicted of murder (later reduced to manslaughter) for the killing of David Wilkie, a taxi driver driving a non-striking miner to work in South Wales during the strike.[43]

Benn stood for election as Party Leader in 1988, against Neil Kinnock, following Labour's third successive defeat in the 1987 general election, and lost by a substantial margin.[44] During the Gulf War, he visited Baghdad to persuade Saddam Hussein to release the hostages who had been captured.[45] He was one of the very few MPs to oppose the Kosovo War.

In 1991, with Labour still in opposition and a general election due by June 1992, he proposed the Commonwealth of Britain Bill, abolishing the Monarchy in favour of the United Kingdom becoming a "democratic, federal and secular commonwealth", a republic with a written constitution. It was read in Parliament a number of times until his retirement at the 2001 election, but never achieved a second reading.[46] He presented an account of his proposal in *Common Sense: A New Constitution for Britain*.[47]

Retirement

Benn did not stand at the 2001 general election; as he explained it, he was "leaving parliament in order to spend more time on politics".[48] Along with Edward Heath, Benn was given the privilege of being able to continue using the House of Commons Library and Members' refreshment facilities by the Speaker. Shortly after his retirement, he was approached by the Stop the War Coalition, and was asked to become its President, an offer he accepted.[45] He thus became a leading figure of the British opposition to the War on Iraq, and in February 2003 he travelled to Baghdad to meet Saddam Hussein. The interview was shown on British television.[49] He spoke against the war at the February 2003 protest in London organised by the Stop the War Coalition, attended by over 1 million people.[50] In February 2004 and 2008, he was re-elected President of the Stop the War Coalition.[51]

Benn about to join the March 2005 anti-war demonstration in London

He has toured with a one-man stage show and appears a few times each year in a two-man show with folk singer Roy Bailey. In 2003, his show with Bailey was voted 'Best Live Act' at the BBC Radio 2 Folk Awards.[52] In 2002 he opened the "Left Field" stage at the Glastonbury Festival. In October 2003, he was a guest of British Airways on the last scheduled Concorde flight from New York to London.[53] In June 2005, Benn was a panellist on a special edition of BBC1's *Question Time* edited entirely by a school-age film crew selected by a BBC competition.[54]

On 21 June 2005, Benn presented a programme on democracy as part of the Channel 5 series *Big Ideas That Changed The World*. He presented a left-wing view of democracy as the means to pass power from the "wallet to the ballot". He argued that traditional social democratic values were under threat in an increasingly globalised world in which powerful institutions such as the International Monetary Fund, the World Bank and the European Commission are unelected and unaccountable to those whose lives they affect daily.[55]

Tony Benn and Giles Fraser speaking at Levellers' Day, Burford, 17 May 2008

On 27 September 2005, Benn became ill while at the Labour Party Conference in Brighton and was taken by ambulance to the Royal Sussex County Hospital after being treated by paramedics at the Brighton Centre. Benn reportedly fell and struck his head. He was kept in hospital for observation and was described as being in a "comfortable condition".[56] He was subsequently fitted with an artificial pacemaker to help regulate his heartbeat.[57] In a list compiled by the magazine *New Statesman* in 2006, he was voted twelfth in the list of "Heroes of our Time".[58]

In September 2006, Benn joined the "Time to Go" demonstration in Manchester the day before the start of the final Labour Party Conference with Tony Blair as Party Leader, with the aim of persuading the Labour Government to withdraw troops from Iraq, to refrain from attacking Iran and to reject replacing the Trident missile and submarines with a new system. He spoke to the demonstrators in the rally afterwards along with other politicians and journalists, including George Galloway and members of the Campaign for Nuclear Disarmament.[59] In 2007, he appeared in an extended segment in the Michael Moore film *Sicko* giving comments about democracy, social responsibility, and health care. A poll by the BBC2 *The Daily Politics* programme in January 2007 selected Benn as the UK's "Political Hero" with 38% of the vote, beating Margaret Thatcher with 35% and five other contenders: Alex Salmond, Leader of the Scottish National Party; Clare Short, Independent MP; Neil Kinnock, previous Labour Party Leader; Norman Tebbit, previous Conservative Party Chairman; and Shirley Williams, one of the 'gang of four' who founded the Social Democratic Party.[5]

In the 2007 Labour Party leadership election, Benn backed the left-wing MP John McDonnell in his unsuccessful bid. In September 2007, Benn called for the government to hold a referendum on the EU Reform Treaty.[60] In October 2007, at the age of 82, and when it appeared that a general election was about to be held, Benn reportedly announced that he wanted to stand, having written to his local Kensington and Chelsea Constituency Labour Party offering himself as a prospective candidate for the seat held by the Conservative Malcolm Rifkind.[61] [62] However, there was no election in 2007, and the Kensington and Chelsea seat was abolished.

In September 2008, Benn appeared on the DVD release for the *Doctor Who* story *The War Machines* with a vignette discussing the Post Office Tower; he became the second Labour politician, after Roy Hattersley to appear in a *Doctor Who* DVD.[63] Also in 2008, Benn appeared on track 12 "Pay Attention to the Human" on Colin MacIntyre's *The Water* album.[64]

Benn on the cover of Dartford Living, Sept 2009

At the Stop the War Conference 2009, he described the wars in Iraq and Afghanistan as "Imperialist war(s)" and discussed the killing of American and allied troops by Iraqi or foreign insurgents, questioning whether they were in fact freedom fighters, and comparing the insurgents to a British Dad's Army, saying "If you are invaded you have a right to self defence, and this idea that people in Iraq and Afghanistan who are resisting the invasion are militant Muslim extremists is a complete bloody lie. I joined Dad's Army when I was sixteen, and if the Germans had arrived, I tell you, I could use a bayonet, a rifle, a revolver, and if I'd seen a German officer having a meal I'd have tossed a grenade through the window. Would I have been a freedom fighter or a terrorist?"[65]

In an interview published in Dartford Living in September 2009, Benn was critical of the Government's decision to delay the findings of the Iraq War Inquiry until after the General Election, stating that "people can take into account what the inquiry has reported on but they've deliberately pushed it beyond the election. Government is responsible for explaining what it has done and I don't think we were told the truth."[66] He also stated that local government was strangled by Margaret Thatcher and had not been liberated by New Labour.[66]

In 2009 Benn was admitted to hospital and *An Evening with Tony Benn*, scheduled to take place at London's Cadogan Hall, was cancelled.[67] He resumed a tour of these shows in 2010. He performed his show *The Writing on the Wall* with Roy Bailey at St Mary's Church, Ashford, Kent in September 2011 as part of the arts venue's first Revelation St Mary's Season.[68] In July 2011 he was awarded an honorary doctorate from the University of Glamorgan, Wales.[69]

In November 2011 it was reported that Benn had moved out of his home in Holland Park Avenue, London, into a smaller £750,000 flat nearby which benefits from a warden.[70]

Diaries and biographies

Tony Benn speaking at the Tolpuddle Martyrs' Festival and Rally 2012

Benn has been a prolific diarist: eight volumes of his diaries have been published (the first six collected as ISBN 0-09-963411-2, the penultimate available as ISBN 0-09-941502-X). He has now stopped writing them following a stroke, and in April 2012 was reportedly working on the final volume.[71] Collections of his speeches and writings were published as *Arguments for Socialism* (1979), *Arguments for Democracy* (1981), (both edited by Chris Mullin), *Fighting Back* (1988) and (with Andrew Hood) *Common Sense* (1993), as well as *Free Radical: New Century Essays* (2004). In August 2003, London DJ Charles Bailey created an album of Benn's speeches (ISBN 1-904734-03-0) set to ambient groove.

He has made public several episodes of audio diaries he made during his time in Parliament and after retirement, entitled 'The Benn Tapes', broadcast originally on BBC Radio 4. Short series have been played periodically on BBC Radio 7.[72] A major biography was written by Jad Adams and published by Macmillan in 1992; it was updated to cover the intervening 20 years and reissued by Biteback Publishing in 2011: *Tony Benn: A Biography* (ISBN 0-333-52558-2) A more recent 'semi-authorised' biography with a foreword by Benn was published in 2001: David Powell, *Tony Benn: A Political Life*, Continuum Books (ISBN 978-0826464156). An autobiography, *Dare to be a Daniel: Then and Now*, Hutchinson (ISBN 978-0099471530), was published in 2004.

There are substantial essays on Benn in the *Dictionary of Labour Biography* by Phillip Whitehead, Greg Rosen [ed], Politicos Publishing, 2001 (ISBN 978-1902301181) and in *Labour Forces: From Ernie Bevin to Gordon Brown*, Kevin Jefferys [ed], I. B. Taurus Publishing, 2002 (ISBN 978-1860647437). Michael Moore dedicates his book *Mike's Election Guide 2008* (ISBN 978-0141039817) to Benn with: "For Tony Benn, keep teaching us".[73]

Plaques

During his final years in Parliament, Benn placed three plaques within the Houses of Parliament. Two are in the 'Suffragettes room' (next to the Central Lobby). The first was placed in 1995 and reads as follows:

"This plaque is dedicated/ with respect, gratitude and affection/ to the many hundreds of thousands of men and women/ who lived and worked on these islands/ and who devoted themselves to the advancement of/ freedom, civil liberties, social justice and democracy/ who campaigned for popular representation in Parliament

Including: Wat Tyler, John Ball, William Tyndale, Thomas More, the Levellers, John Lilburne, William Walwyn, The Diggers, Gerard Winstanley, Tom Paine, Mary Wollstonecraft, Robert Owen, The Tolpuddle Martyrs, the Chartists, Keir Hardie, Annie Besant, Charles Bradlaugh, Robert Tressell, the Suffragettes and *Constance Markievicz*

And many others whose names have never/ been recorded in our history"

The second was placed in 1996 and is dedicated to all who work within the Houses of Parliament.

The third is dedicated to Suffragette Emily Wilding-Davison and was placed in the broom cupboard next to the Undercroft Chapel within the Palace of Westminster, where Davison is said to have hidden during the 1911 census in order to establish her address as the House of Commons.[74]

He unveiled a plaque in Highbury, North London, to commemorate the Peasants' Revolt of 1381.[75]

See also

- Labour Representation Committee
- Socialist Campaign Group

References

[1] Tony Benn (2007). *More Time for Politics: Diaries 2001–2007*. London: Hutchinson. p. 223.

[2] http://www.tonybenn.com/

[3] *Re Parliamentary Election for Bristol South East* [1964] 2 Q.B. 257, [1961] 3 W.L.R. 577

[4] Renton, Dave (February 1997). "Does Labour's Left Have an Alternative?" (http://pubs.socialistreviewindex.org.uk/sr205/labour.htm). *Socialist Review*. . Retrieved 2 May 2010.

[5] "The Magnificent Seven political heroes..." (http://news.bbc.co.uk/1/hi/programmes/politics_show/6161847.stm). *BBC News* (BBC). 12 December 2006. . Retrieved 4 October 2007.

[6] "Collection – The Rt Hon Tony Benn MP –" (http://art.parliament.uk/worksofart/artwork/andrew-tift/the-rt-hon-tony-benn-mp/4761/). *Art in Parliament*. UK Parliament. . Retrieved 3 May 2010.

[7] "Tony Benn – Official Website" (http://www.tonybenn.com/Biography.html). *tonybenn.com*. . Retrieved 2 May 2010.

[8] Hale, Leslie (January 2008). "Benn, William Wedgwood" (http://www.oxforddnb.com/view/article/30705?docPos=2) (Subscription required). *Oxford Dictionary of National Biography Online*. Oxford University Press. . Retrieved 2 May 2010.

[9] Brodie, Marc (January 2008). "Benn, Sir John Williams" (http://www.oxforddnb.com/view/article/58236/?back=,30705) (Subscription required). *Oxford National Dictionary of Biography Online*. Oxford University Press. . Retrieved 2 May 2010.

[10] Stearn, Roger T (2004). "Benn, Margaret Eadie Wedgwood" (http://www.oxforddnb.com/view/article/50714/?back=,30705) (Subscription required). *Oxford National Dictionary of Biography Online*. Oxford University Press. . Retrieved 2 May 2010.

[11] "Tony Benn: You Ask The Questions" (http://www.independent.co.uk/news/people/profiles/tony-benn-you-ask-the-questions-503487.html). *The Independent* (London: Independent News and Media). 19 August 2005. . Retrieved 2 May 2010.

[12] Benn, Tony (2003). *Free Radical* (http://books.google.com/?id=UsYKitlTrdYC&printsec=frontcover&dq=tony+benn+free+radical&q). Continuum. p. 226. ISBN 0-8264-6596-X. .

[13] "Mr Benn wipes away his past". *The Times Diary* (Times Newspapers). 18 March 1976.

[14] "Not Out". *The Times Diary* (Times newspapers). 4 April 1977.

[15] "Tony Benn" (http://www.thebiographychannel.co.uk/biography_story/455:0/1/Tony_Benn.htm). *The Biography Channel*. . Retrieved 2 April 2007.

[16] "William Wedgwood Benn" (http://www.spartacus.schoolnet.co.uk/TUbenn.htm). *Spartacus Educational*. . Retrieved 2 April 2007.

[17] "Caroline Benn" (http://www.telegraph.co.uk/news/obituaries/1375544/Caroline-Benn.html). *Daily Telegraph* (London: Telegraph Media Group). 24 November 2000. . Retrieved 2 May 2010.

[18] "Benn's granddaughter runs for MP" (http://news.bbc.co.uk/1/hi/uk_politics/7011922.stm). *BBC News* (BBC). 25 September 2007. . Retrieved 2 May 2010.

[19] Skinitis, Alexia (10 January 2009). "Emily Benn the younger" (http://www.timesonline.co.uk/tol/news/politics/article5452741.ece). *Times Online* (London: Times Newspapers). . Retrieved 2 May 2010.

[20] French, Philip (26 July 2009). "Philip French's screen legends: Margaret Rutherford" (http://www.guardian.co.uk/film/2009/jul/26/screen-legends-margaret-rutherford). *Guardian News and Media* (London). . Retrieved 2 May 2010.

[21] "The Benn dynasty" (http://news.bbc.co.uk/1/hi/uk_politics/366619.stm). *BBC News* (BBC). 11 June 1999. . Retrieved 2 May 2010.

[22] "New Members Sworn" (http://hansard.millbanksystems.com/commons/1950/dec/04/new-members-sworn#S5CV0482P0_19501204_HOC_256). *Hansard*. House of Commons. 4 December 1950. . Retrieved 2 May 2010.

[23] "New Members Sworn" (http://hansard.millbanksystems.com/commons/1950/dec/05/new-member-sworn#S5CV0482P0_19501205_HOC_264). *Hansard*. House of Commons. 5 December 1950. . Retrieved 2 May 2010.

[24] "Profile: Tony Benn" (http://www.bbc.co.uk/bristol/content/articles/2008/04/03/tony_benn_profile_feature.shtml). *BBC Bristol*. BBC. 4 April 2008. . Retrieved 2 May 2010.

[25] "Disclaiming a peerage" (http://news.bbc.co.uk/1/hi/uk_politics/82121.stm). *BBC News* (London: British Broadcasting Corporation). 14 July 2005. . Retrieved 7 June 2008.

[26] "A Revolution in British Stamp Design" (http://postalheritage.org.uk/exhibitions/onlineexhibitions/gentleman-on-stamps/the-david-gentleman-album). *The British Postal Museum and Archive*. . Retrieved 2 May 2010.

[27] Lynskey, Dorian (27 March 2009). "All bands on deck: Director Richard Curtis and DJ Johnnie Walker on the making of The Boat That Rocked" (http://www.guardian.co.uk/film/2009/mar/27/boat-that-rocked-background). *The Guardian* (London: Guardian News and Media). . Retrieved 2 May 2010.

[28] "UK Confidential Transcripts: Tony Benn – The Labour Minister" (http://news.bbc.co.uk/1/hi/programmes/uk_confidential/1706972.stm). *BBC News* (BBC). 1 January 2002. . Retrieved 2 May 2010.

[29] Benn, Tony (1995). *The Benn Diaries*. Arrow. p. 432. ISBN 978-0-09-963411-9.

[30] Powell, David (2003). *Tony Benn: a political life* (http://books.google.com/?id=dYeVIfy0by8C&pg=PA82&lpg=PA82&dq=alternative+economic+strategy+tony+benn&q=alternative economic strategy tony benn) (2 ed.). London & New York: Continuum. pp. 82, 84. ISBN 0-8264-7074-2. .

[31] Benn, Tony (1988). *Out of the Wilderness: Diaries 1963–67*. Arrow. p. Introduction. ISBN 978-0-09-958670-8.

[32] Kavanagh, Dennis (1990). "Tony Benn: Nuisance or Conscience?". In Kavanagh, Dennis. *Politics and Personalities*. Macmillan. p. 184. ISBN 978-0-333-51580-8.

[33] Kavanagh, Dennis (1990). "Tony Benn: Nuisance or Conscience?". In Kavanagh, Dennis. *Politics and Personalities*. Macmillan. p. 178. ISBN 978-0-333-51580-8.

[34] Whiteley, Paul (11 January 1980). "The Labour Party: Middle Class, Militant and Male". *New Statesman*: 41–42.

[35] "Benn's call for SF to take seats" (http://news.bbc.co.uk/1/hi/uk_politics/northern_ireland/4540343.stm). *BBC News*. 12 May 2005. . Retrieved 2 May 2010.

[36] Emery, Fred (30 September 1980). "Mr Benn proposes timetable of one month to abolish Lords and leave EEC" (http://infotrac.galegroup.com/itw/infomark/27/443/107065937w16/purl=rc1_TTDA_0_CS17008446&dyn=12!xrn_20_0_CS17008446&hst_1?sw_aep=uwesteng) (Subscription required). *The Times, archived by Gale Group* (Times Newspapers). . Retrieved 3 May 2010.

[37] Taylor, Gary (February 2001). "Media Review: The Transformation Of Labour" (http://findarticles.com/p/articles/mi_m2242/is_1621_278/ai_71712189/pg_2/?tag=content;col1). *Contemporary Review* (findarticle.com): 2–4. . Retrieved 3 May 2010.

[38] "House of Commons Statement: Falkland Islands" (http://www.margaretthatcher.org/speeches/displaydocument.asp?docid=104969). Margaret Thatcher Foundation. 15 June 1982. . Retrieved 4 October 2007.

[39] "Benn on the couch". *The Sun* (News international). 1 March 1984.

[40] "Labour's new line-up" (http://infotrac.galegroup.com/itw/infomark/27/443/107065937w16/purl=rc1_TTDA_0_CS51350883&dyn=21!xrn_30_0_CS51350883&hst_1?sw_aep=uwesteng) (Subscription required). *The Times, archived by Gale Group* (Times Newspapers). 3 November 1983. . Retrieved 3 May 2010.

[41] "Chapter 06; ...1974 strike...a conversation with miners...Labour government... ...Benn helps divide miners..." (http://libcom.org/library/chapter-06-1974-strikea-conversation-minerslabour-government-benn-helps-divide-miners). *libcom.org*. . Retrieved 3 May 2010.

[42] Robertson, Jack (23 April 2010). "25 years after the Great Miners' Strike" (http://www.isj.org.uk/index.php4?id=640&issue=126). *International Socialism* (London: Socialist Workers Party) (126). . Retrieved 3 May 2010.

[43] "Miners' Amnesty (General Pardon)" (http://hansard.millbanksystems.com/commons/1985/jun/28/miners-amnesty-general-pardon). *Hansard*. House of Commons. 28 June 1985. . Retrieved 2 May 2010.

[44] Durkin, Mary. "Leadership Elections: Labour Party" (http://www.parliament.uk/commons/lib/research/briefings/snpc-03938.pdf) (PDF). *House of Commons Library*. . Retrieved 3 May 2010.

[45] Stadlen, Nick (8 December 2006). "Brief Encounter: Tony Benn" (http://www.guardian.co.uk/politics/2006/dec/08/1). *The Guardian* (London: Guardian News and Media). . Retrieved 3 May 2010.

[46] "Commonwealth of Britain Bill" (http://hansard.millbanksystems.com/bills/commonwealth-of-britain-bill). *Hansard*. House of Commons. . Retrieved 3 May 2010.

[47] Benn, Tony; Hood, Andrew (17 June 1993). Hood, Andrew. ed. *Common Sense: New Constitution for Britain*. London: Hutchinson. ISBN 978-0-09-177308-3.

[48] Gary Younge "The stirrer" (http://www.guardian.co.uk/politics/2002/jul/20/interviews.labour), *The Guardian*, 20 July 2002

[49] "Full text of Benn interview with Saddam" (http://news.bbc.co.uk/1/hi/2726831.stm). *BBC News*. 4 February 2003. . Retrieved 3 May 2010.

[50] "'Million' march against Iraq war" (http://news.bbc.co.uk/1/hi/2765041.stm). *BBC News*. 16 February 2003. . Retrieved 3 May 2010.

[51] "Elected positions" (http://web.archive.org/web/20100401145946/http://stopwar.org.uk/content/view/20/52/). *Stop the War Coalition*. Archived from the original (http://stopwar.org.uk/content/view/20/52/) on 1 April 2010. . Retrieved 3 May 2010.

[52] "Promoters' Pack – Roy Bailey & Tony Benn" (http://www.roybailey.net/promotersPack/rbtb.php?PHPSESSID=50526581978a2d6f9d2fc747533000d8). *Roy Bailey*. . Retrieved 3 May 2010.

[53] "Concorde Aircraft Facts, Dates and History" (http://www.flightlevel350.com/Concorde_aircraft_facts.html). *flightlevel350.com*. . Retrieved 3 May 2010.

[54] "Question Time : A question of citizenship" (http://news.bbc.co.uk/1/hi/programmes/question_time/4642035.stm). *BBC News* (BBC). 1 July 2005. . Retrieved 3 May 2010.

[55] Wardle, Tony (Spring 2006). "The Extraordinary Adventures of Mr Benn" (http://www.viva.org.uk/celebs/tonybenn_interview.html). *Viva!LIFE* (Viva!) (31). . Retrieved 2 May 2010.

[56] Staff (28 September 2005). "Tony Benn 'comfortable' in hospital after fall" (http://www.guardian.co.uk/politics/2005/sep/28/labourconference.labour2). *Guardian News and Media* (London). . Retrieved 3 May 2010.

[57] "Benn gets pacemaker after fall" (http://news.bbc.co.uk/1/hi/uk_politics/4300306.stm). *BBC NEWS* (BBC). 1 October 2005. . Retrieved 3 May 2010.

[58] Cowley, Jason (22 May 2006). "Heroes of our time – the top 50" (http://www.newstatesman.com/200605220016). New Statesman. . Retrieved 4 October 2007.

[59] "Thousands at city's anti-war demo" (http://news.bbc.co.uk/1/hi/england/manchester/5373128.stm). *BBC News*. 23 September 2006. . Retrieved 3 May 2010.

[60] "Give us EU referendum, says Benn" (http://news.bbc.co.uk/1/hi/uk_politics/7011231.stm). *BBC News* (BBC). 24 September 2007. . Retrieved 2 May 2010.

[61] "I want to be an MP again – Benn" (http://news.bbc.co.uk/1/hi/uk_politics/7028042.stm). *BBC News* (BBC). 4 October 2007. . Retrieved 4 October 2007.

[62] Attewill, Fred (4 October 2007). "Benn: I want to return to parliament" (http://www.guardian.co.uk/uk_news/story/0,,2183661,00.html). *The Guardian* (Guardian News and Media). . Retrieved 5 October 2007.

[63] Wilkins, Jonathan (21 August 2008). "Doctor Who: The War Machines Review" (http://totalscifionline.com/reviews/2386-doctor-who-the-war-machines). *Total SciFi Online*. . Retrieved 3 May 2010.

[64] Sandall, Robert (7 February 2008). "Tony Benn, pop star" (http://www.telegraph.co.uk/culture/music/3670995/Tony-Benn-pop-star.html). *Telegraph Media Group* (London). . Retrieved 3 May 2010.

[65] Jesse Oldershaw (camera); Andy Cousins (editor) (25 April 2009) (Adobe Flash). *Tony Benn – Stop the War Conference 2009* (http://www.youtube.com/watch?v=og4K4yhQF8E) (Streaming http). Stop the War Coalition. Event occurs at 3:06. .

[66] Khairoun, Abdel (September 2009). "Big Benn Chimes in to Dartford" (http://www.dartfordadvertising.co.uk/uploaded_files/Dartford Living September 2009.pdf) (PDF). *Dartford Living*. p. 10. . Retrieved 2 May 2010.

[67] "Cadogan Hall : Featured Event" (http://www.cadoganhall.com/showpage.php?pid=902). *cadoganhall.com*. 12 October 2009. . Retrieved 7 July 2011.

[68] "Clive Conway Celebrity Productions – An Audience with an Evening With Tony Benn" (http://www.celebrityproductions.info/displayer_productions.php/178/An_Evening_with_Tony_Benn). *celebrityproductions.info*. 2011 [last update]. . Retrieved 7 July 2011.

[69] "University of Glamorgan honours contributions to public life, communities, science, literature, and sport" (http://news.glam.ac.uk/news/en/2011/jun/28/university-glamorgan-honours-contributions-public-/). *news.glam.ac.uk*. 2011 [last update]. . Retrieved 7 July 2011.

[70] Richard Kay "Frail Tony Benn downsizes to £750,000 flat (well, his house was worth £3m)", (http://www.dailymail.co.uk/news/article-2055900/RICHARD-KAY-Frail-Tony-Benn-downsizes-750-000-flat-house-worth-3m.html) *Daily Mail*, 1 November 2011. Retrieved 5 November 2011.

[71] "Happy Birthday Tony Benn, 87" (http://www.thetimes.co.uk/tto/life/courtsocial/article3371918.ece), *The Times*, 3 April 2012

[72] "BBC Radio 7 Programmes – The Benn Tapes" (http://www.bbc.co.uk/programmes/b007jr2r#synopsis). *BBC*. 16 March 2009. . Retrieved 3 May 2010.

[73] Goodman, Amy (1 November 2008). "Michael Moore on the Election, the Bailout, Healthcare, and 10 Proposals for the Next President by Michael Moore" (http://www.zcommunications.org/michael-moore-on-the-election-the-bailout-healthcare-and-10-proposals-for-the-next-president-by-michael-moore). *Democracy Now!*. ZCommunications. . Retrieved 3 May 2010.

[74] "Benn's secret tribute to suffragette martyr" (http://entertainment.timesonline.co.uk/tol/arts_and_entertainment/music/article6529773.ece). *BBC News*. 17 March 1999. . Retrieved 3 November 2011.

[75] "Tony Benn to unveil Islington People's Plaque commemorating the Peasants' Revolt" (http://www.islington.gov.uk/Council/CouncilNews/PressOffice/2011/06/PR4430.asp). *Islington Borough Council*. 2 June 2011. . Retrieved 3 November 2011.

Bibliography

- *Speeches*, Spokesman Books (1974) ISBN 0851240917
- *Levellers and the English Democratic Tradition*, Spokesman Books (1976) ISBN 978-0-85124-633-8
- *Why America Needs Democratic Socialism*, Spokesman Books (1978) ISBN 978-0-85124-266-8
- *Prospects*, Amalgamated Union of Engineering Workers, Technical, Administrative and Supervisory Section (1979)
- *Case for Constitutional Civil Service*, Inst. for Workers' Control (1980) ISBN 978-0-901740-67-0
- *Case for Party Democracy*, Inst. for Workers' Control (1980) ISBN 978-0-901740-70-0
- *Arguments for Socialism*, Penguin Books (1980) ISBN 978-0-14-005489-7
- & Chris Mullin, *Arguments for Democracy*, Jonathan Cape (1981) ISBN 978-0-224-01878-4
- *European Unity: A New Perspective*, Spokesman Books (1981) ISBN 978-0-85124-326-9
- *Parliament and Power: Agenda for a Free Society*, Verso Books (1982) ISBN 978-0-86091-057-2
- & Andrew Hood, *Common Sense: New Constitution for Britain*, Hutchinson (1993)
- *Free Radical: New Century Essays*, Continuum International Publishing (2004) ISBN 978-0-8264-7400-1

- *Dare to Be a Daniel: Then and Now*, Hutchinson (2004) ISBN 978-0-09-179999-1
- *Letters to my Grandchildren: Thoughts on the Future*, Arrow Books (2010) ISBN 978-0-09-953909-4

Diaries

- *Out of the Wilderness: Diaries 1963–67*, Hutchinson (1987) ISBN 978-0-09-170660-9
- *Office Without Power: Diaries 1968–72*, Hutchinson (1988) ISBN 978-0-09-173647-7
- *Against the Tide: Diaries 1973–76*, Hutchinson (1989) ISBN 978-0-09-173775-7
- *Conflicts of Interest: Diaries 1977–80*, Hutchinson (1990) ISBN 978-0-09-174321-5
- *The End of an Era: Diaries 1980–90*, Hutchinson (1992) ISBN 978-0-09-174857-9
- *Years of Hope: Diaries 1940–62*, Hutchinson (1994) ISBN 978-0-09-178534-5
- *The Benn Diaries: Single Volume Edition 1940–90*, Hutchinson (1995) ISBN 978-0-09-179223-7
- *Free at Last!: Diaries 1991–2001*, Hutchinson (2002) ISBN 978-0-09-179352-4
- *More Time for Politics: Diaries 2001–2007*, Hutchinson (2007) ISBN 978-0-09-951705-4

External links

- Hansard 1803–2005: contributions in Parliament by Tony Benn (http://hansard.millbanksystems.com/people/mr-tony-benn)
- Unofficial Tony Benn site with extensive interview, articles, audio and visual (http://www.bennites.com/)
- 'Face-to-Face with Tony Benn'. Freeview video interview by the Vega Science Trust (http://www.vega.org.uk/video/programme/26)
- Unofficial Tony Benn Quotation site (http://tony-benn.blogspot.co.uk/)
- Tony Benn's interview with *The Third Estate*. (http://thethirdestate.net/2009/07/an-interview-with-tony-benn/)
- Andrew Roth. Tony Benn Chesterfield and Bristol South East MP (http://politics.guardian.co.uk/profiles/story/0,9396,-361,00.html), *The Guardian*, 25 March 2001.
- *Guardian* web guide to the veteran leftwinger Tony Benn (http://www.guardian.co.uk/netnotes/article/0,6729,728418,00.html)
- Tony Benn interview for Leftfield appearances (http://www.glastonburyfestivals.co.uk/news/index.asp?id=212) at the Glastonbury Festival
- Tony Benn. Atomic hypocrisy: West is not in a position to take a high moral line (http://politics.guardian.co.uk/publicservices/comment/0,11032,1653785,00.html), *The Guardian*, 30 November 2005.
- Amy Goodman. Interview with Tony Benn: How Britain secretly helped Israel build its nuclear arsenal (http://www.democracynow.org/article.pl?sid=06/03/10/1451237), *Democracy Now!*, 10 March 2006.
- Benn in the 1960s (http://ugandandiscussions.co.uk/covers/147_big.jpg) as seen by *Private Eye*
- Audio interview with *The Guardian*. (http://blogs.guardian.co.uk/podcasts/2006/12/brief_encounter_tony_benn.html)
- His Address (http://www.thehist.com/index.php?option=com_docman&task=cat_view&gid=80&Itemid=455) to the College Historical Society of Trinity College
- Interview with Tony Benn – Radio France Internationale (http://www.rfi.fr/actuen/articles/111/article_3308.asp) – 28 March 2008 – 6-minute audio – Ahead of G20 marches, London
- Tony Benn on Tony Blair: "He Is Guilty of a War Crime" (http://www.democracynow.org/2010/9/21/british_mp_tony_benn_on_tony) – video report by *Democracy Now!*

Charles_Longley

Charles Longley[1]	
Archbishop of Canterbury	

Enthroned	1862
Reign ended	27 October 1868
Predecessor	John Sumner
Successor	Archibald Tait
Personal details	
Born	1794 Rochester, Kent
Died	27 October 1868 Addington Palace
Buried	St Mary the Blessed Virgin Church, Addington, London

Charles Thomas Longley (28 July 1794 – 27 October 1868[2]) was a bishop in the Church of England. He served as Bishop of Ripon, Bishop of Durham, Archbishop of York and Archbishop of Canterbury from 1862 until his death.

Life

He was born at Rochester, the fifth son of the late John Longley, Recorder of Rochester,[3] and educated at Westminster School and the University of Oxford. He was ordained in 1818, and was appointed vicar of Cowley, Oxford, in 1823. In 1827, he received the rectory of West Tytherley, Hampshire, and two years later he was elected headmaster of Harrow School. He held this office until 1836, when he was consecrated bishop of the new see of Ripon. In 1856 he became Bishop of Durham, and in 1860 he became Archbishop of York.

In 1862, he succeeded John Bird Sumner as Archbishop of Canterbury. Soon afterwards the questions connected with the deposition of Bishop John William Colenso were referred to Longley but, while regarding Colenso's opinions as heretical and his deposition as justifiable, he refused to pronounce upon the legal difficulties of the case.

The chief event of his primacy was the meeting at Lambeth, in 1867, of the first Pan-Anglican conference of British, colonial and foreign bishops. His published works included numerous sermons and addresses. He died at Addington Park, near Croydon.

Family

As Headmaster of Harrow School, he married 15 December 1831 Caroline Sophia Parnell, whose brother Hon and Rev George Damer Parnell MA was the curate of Ash 1859–1861. Miss Parnell was the daughter of Sir Henry Brooke Parnell 4th Baronet, cr. (1841) first Baron Congleton by his wife Lady Caroline Elizabeth Dawson, the eldest daughter of John, first Earl of Portarlington.[4] They had several children, three sons and three daughters,[5] of whom

A photograph of Charles Thomas Longley by Charles Dodgson (Lewis Carroll).

1. Sir Henry Longley, KCB (28 November 1833 – 25 December 1899), served as Chief Charity Commissioner for the British government.[6] He married 17 September 1861 Diana Eliza Davenport (fl. 1905), daughter of John Davenport of Foxley, Herefordshire.[7] 1.1. John Augustine Longley, served as assistant private secretary to the Lord Privy Seal and married 26 May 1898 Lady Louisa Katherine Scott, daughter of the 3rd Earl of Eldon. There was no issue listed for the marriage, nor a birthdate for John Augustine Longley.

Two younger sons died apparently unmarried.

4. Mary Henrietta Longley married 9 December 1858 the Hon. Rev. George Winfield Bourke (d. 9 October 1903), Honorary Chaplain to the Monarch, and son of Robert Bourke, 5th Earl of Mayo. Their only child Walter Longley Bourke (b. 28 November 1859; d. 1939) became the 8th Earl of Mayo); from 1891 to 1903, he was a Trustee of the Bridgewater Estates. He had married in 1887, and had issue, four sons and two daughters, by 1905.[8] The second son Ulick Henry Bourke (1890–1962) became 9th Earl of Mayo, and third son Hon Bryan Longley Bourke (1897–1961) was father of the 10th Earl of Mayo (1929–2006), father of the present Earl.

5. Caroline Georgina Longley (d. 30 October 1867) married 6 November 1862[9] (as his 1st wife) Major Edward Levett (18 December 1832 – 28 December 1899), 10th Royal Hussars, of Wychnor Park and Packington Hall, Staffordshire, third son of John Levett and his wife Sophia Kennedy, granddaughter of Archibald Kennedy, 11th Earl of Cassilis. They had issue 2 daughters, both of whom married and had issue[10]

Charles Thomas Longley, by George Richmond, c. 1862

6. Rosamond Esther Harriett Longley (d. 1936) married 1870 Hon. Cecil Thomas Parker (1845–1931), 2nd son of the 6th Earl of Macclesfield by his 2nd wife Lady Mary Frances Grosvenor, a sister of the 1st Duke of Westminster, and had issue 4 sons and 2 daus. Their elder daughter Caroline Beatrix Parker, later Viscountess Bridgeman DBE (1875–1861) married 1895 William Clive Bridgeman, who became The Rt. Hon. 1st Viscount Bridgeman, of Leigh, Shropshire in 1929, PC (1864–1935), only child of Rev. Hon. John Orlando Bridgeman, Rector of Weston-under-Lizard (himself 3rd and youngest son of George Bridgeman, 2nd Earl of Bradford) by his wife Marianne Caroline Clive, daughter of Ven. William Clive, and left issue, including the present Viscount.[11] The fourth and youngest son Wilfrid Parker (1883–1966) became the Rt.Rev. Hon. Bishop of Pretoria, South Africa. A granddaughter (by the 3rd son Geoffrey) Isolda Rosamond Parker (1918–?) married 1940 David Bertram Pollock, 2nd Viscount Hanworth (1916–1996) and is mother of the present peer.[12]

Notes

[1] Archbishop of Canterbury: Better Bishops for the sake of a better Church (http://www.lambethconference.org/lc2008/news/news.cfm/2008/4/23/Archbishop-of-Canterbury-Better-Bishops-for-the-sake-of-a-better-Church)

[2] "J. R. Garrard, 'Longley, Charles Thomas (1794–1868)', *[[Oxford Dictionary of National Biography* (http://www.oxforddnb.com/view/article/16987)], *Oxford University Press, 2004"]. . Retrieved 6 June 2008.*

[3] "Obituary: Charles Thomas Longley, D.D., Archbishop of Canterbury" (http://query.nytimes.com/mem/archive-free/pdf?_r=1&res=9B06E1D61330EE34BC4151DFB6678383679FDE) *The New York Times*, 29 October 1868, p. 4, (citation only (http://query.nytimes.com/gst/abstract.html?res=9B06E1D61330EE34BC4151DFB6678383679FDE)). Retrieved 3 December 2008

[4] Leslie Morgan. "A Victorian Curate of Ash and his Brother-in-law's Letter" (http://www.ash-church.org.uk/lclhis4.php) St Peter and St Paul, Ash Church Website. Portarlington is misspelled Porterlington. Retrieved 3 December 2008

[5] Marquis of Ruvigny & Raineval The Plantagenet Roll of the Blood Royal: The Clarence Volume, Containing the Descendants of George, Duke of Clarence (http://books.google.co.in/books?id=w9UYYThhRIQC&pg=PA287&lpg=PA287&dq="Charles+Thomas+Longley"+Parnell&source=bl&ots=mn5p41SaF7&sig=PRtmSmoUYFtZnpkkH90tUBQFDEE&hl=en&sa=X&oi=book_result&resnum=8&ct=result) p. 287. Published by Genealogical Publishing Com, 1994 ISBN 0-8063-1432-X, 9780806314327. Originally published: London : T.C. & E.C. Jack, 1905. All details below are from this source, unless cited otherwise.

[6] Memorial wall tablet in the church of St Mary the Blessed Virgin Mary, Addington, Surrey

[7] Marquis of Ruvigny & Raineval The Plantagenet Roll of the Blood Royal: The Clarence Volume, Containing the Descendants of George, Duke of Clarence (http://books.google.co.in/books?id=w9UYYThhRIQC&pg=PA287&lpg=PA287&dq="Charles+Thomas+Longley"+Parnell&source=bl&ots=mn5p41SaF7&sig=PRtmSmoUYFtZnpkkH90tUBQFDEE&hl=en&sa=X&oi=book_result&resnum=8&ct=result) p. 287.

[8] *Ibid*

[9] The Annual Register, J. & F.H. Rivington, London, 1862 (http://books.google.com/books?id=coNdAAAAIAAJ&pg=RA2-PA320&lpg=RA2-PA320&dq="caroline+georgiana+longley"&source=bl&ots=rfuzIlvg2Y&sig=EeRK-QgH7IgVcqKbnwWpZ9jkwe4&hl=en&ei=QZ2GSq2PDIKysgODp6SeBw&sa=X&oi=book_result&ct=result&resnum=1#v=onepage&q="caroline georgiana longley"&f=false)

[10] Marquis de Ruvigny, Raineval Staff The Plantagenet Roll of the Blood Royal: The Clarence Volume (http://books.google.co.in/books?id=ObnB2K0Pm-gC&pg=PA338&lpg=PA338&dq="Charles+Thomas+Longley"&source=web&ots=UxFjUkbd0p&sig=ky5t2ae-rG8VTh62eJcoXRWnP-0&hl=en&sa=X&oi=book_result&resnum=6&ct=result) and Clarence Volume, p. 287 (http://books.google.co.in/books?id=w9UYYThhRIQC&pg=PA287&lpg=PA287&dq="Charles+Thomas+Longley"+Parnell&source=bl&ots=mn5p41SaF7&sig=PRtmSmoUYFtZnpkkH90tUBQFDEE&hl=en&sa=X&oi=book_result&resnum=8&ct=result), reprinted by Genealogical Publishing Company, 1994. Retrieved 3 December 2008. The date of Caroline Georgina's marriage is not given by Ruvigny, nor is her birthdate.

[11] Clarence Volume, p. 287 (http://books.google.co.in/books?id=w9UYYThhRIQC&pg=PA287&lpg=PA287&dq="Charles+Thomas+Longley"+Parnell&source=bl&ots=mn5p41SaF7&sig=PRtmSmoUYFtZnpkkH90tUBQFDEE&hl=en&sa=X&oi=book_result&resnum=8&ct=result), reprinted by Genealogical Publishing Company, 1994. Retrieved 3 December 2008. Also see Conqueror – William 165 (http://www.william1.co.uk/w165.htm) and Conqueror – William 176 (http://www.william1.co.uk/w176.htm#w176l12). Retrieved 3 December 2008

[12] Conqueror – William 50 to 52 (http://www.william1.co.uk/w505152.htm#w505152l8). Retrieved 3 December 2008

References

This article incorporates text from a publication now in the public domain: Chisholm, Hugh, ed. (1911). *Encyclopædia Britannica* (11th ed.). Cambridge University Press.

- "Obituary: Charles Thomas Longley, D.D., Archbishop of Canterbury" (http://query.nytimes.com/mem/archive-free/pdf?_r=1&res=9B06E1D61330EE34BC4151DFB6678383679FDE) *The New York Times*, 29 October 1868, p. 4, (citation only (http://query.nytimes.com/gst/abstract.html?res=9B06E1D61330EE34BC4151DFB6678383679FDE))

External links

- Bibliographic directory (http://anglicanhistory.org/england/ctlongley/) from Project Canterbury
- Archbishop Longley's papers are held at Lambeth Palace Library (http://archives.lambethpalacelibrary.org.uk:8080/Archives/dserve.exe?dsqIni=Dserve.ini&dsqApp=Archive&dsqCmd=Show.tcl&dsqDb=Catalog&dsqSearch=(RefNo=='longley'))

John_Thurso

The Right Honourable **The Viscount Thurso** MP	
[[file:John Thurso MP at Bournemouth 2009.jpg	alt=]]
Chairman of the Finance and Services Committee	
Incumbent	
Assumed office 6 May 2010	
Preceded by Sir Stuart Bell	
Member of Parliament **for Caithness, Sutherland and Easter Ross**	
Incumbent	
Assumed office 7 June 2001	
Preceded by Robert Maclennan	
Majority 4,826 (16.8%)	
Personal details	
Born 10 September 1953Thurso, Caithness, Scotland	
Nationality British	
Political party Liberal Democrats	
Spouse(s) Marion, Lady Thurso (née Sage)	
Children 1 daughter, 2 sons	
Alma mater Eton College	
Religion Episcopalian	

John Archibald Sinclair, 3rd Viscount Thurso (born 10 September 1953), known as **John Thurso**, is a Scottish businessman and Liberal Democrat politician. He is the Member of Parliament (MP) for Caithness, Sutherland and Easter Ross and the fifth generation of the Sinclair family to represent the Caithness area in the House of Commons.

Education, family and non-political career

John Sinclair was educated in Thurso and at Eton College. Thurso joined the Savoy Group as a management trainee in 1972 and following this worked for many years in the hospitality industry. His achievements included managing the Lancaster in Paris (1981–85) and founding the hotel at Cliveden (1985–92) before becoming CEO of Granfel Holdings, owners of East Sussex National Golf Course from 1992 to 1995. Finally from 1995 until his election to parliament in 2001 he was CEO of the Champneys Group. During his time in this job he featured in the TV documentary *Trouble at the Top - Shape up with Lord Thurso.*

Lord Thurso comes from a family of Liberal politicians. The former constituency of Caithness and Sutherland had been held by his grandfather, Archibald Sinclair from 1922 until 1945. Archibald Sinclair was the 1st Viscount Thurso and a Liberal Party leader. Thurso has been married to Marion for 26 years and they have a daughter and two sons. The family live at Thurso, Caithness.

Thurso holds the Presidency of The Tourism Society[1] and the Academy of Food and Wine Service.[2] He is a Fellow of the HCIMA and served as a Patron for six years, until June 2003. He is President of the British International Spa Association,[3] a Trustee of the Clan Sinclair Trust, as well as being Patron of the Bluebell Railway 50th Anniversary Appeal in which capacity, on 24 April 2009, at the railway's Horsted Keynes station his lordship carried out the ceremonial renaming of the Battle of Britain class locomotive named after his grandfather, Sir Archibald Sinclair, Secretary of State for Air during that battle.[4]

Political career

Following his father's death in 1995 he took his seat in the House of Lords as the 3rd Viscount Thurso where he was spokesman on tourism and later food matters. Thurso spoke many times in the House of Lords in favour of Lords reform. His automatic right as a hereditary peer to sit in the House of Lords was abolished in 1999, and he did not attempt to remain.[5] At the 2001 general election he was elected to the House of Commons to represent Caithness, Sutherland and Easter Ross becoming the first British hereditary peer allowed to sit in the House of Commons without having to disclaim his title.[5] He served as Liberal Democrat Shadow Scotland Secretary under Charles Kennedy,[6] but was sacked by Sir Menzies Campbell. He has publicly gone against party policy by declaring his support for nuclear power,[7] and his criticism of 24-hour drinking and wind power.

See also

- Liberal Democrat Frontbench Team
- Clan Sinclair

References

[1] www.tourismsociety.org (http://www.tourismsociety.org/)
[2] www.afws.co.uk (http://www.afws.co.uk/)
[3] BISA Council and Associates British International Spa Association (http://www.spaassociation.org.uk/bisa/council-and-associates)
[4] www.bulleidsociety.org (http://www.bulleidsociety.org/34059/34059_Renaming.html)
[5] Parkinson, Justin (22 February 2011). "John Thurso: The hereditary peer who became an MP" (http://www.bbc.co.uk/news/uk-politics-12419022). BBC News. . Retrieved 29 December 2011.
[6] Uncredited (3 November 2001). "Kennedy boosts his frontbench team" (http://www.scotsman.com/news/politics/kennedy_boosts_his_frontbench_team_1_582679). *The Scotsman*. . Retrieved 29 December 2011. "One of the notable changes was the promotion of new MP Lord John Thurso (Caithness, Sutherland and Easter Ross) to the Scottish affairs brief."
[7] Edwards, Rob (5 March 2006). "Nuclear power: splitting the LibDems and Labour" (http://pqasb.pqarchiver.com/smgpubs/access/997967111.html?dids=997967111:997967111&FMT=ABS&FMTS=ABS:FT&type=current&date=Mar+05,+2006&author=INVESTIGATION+By+Rob+Edwards+Environment+Editor&pub=Sunday+Herald&desc=Nuclear+power:+splitting+the+LibDems+and+Labour+ENERGY:+FISSIONS+ENERGY:+FISSIONS+Row+over+lobbyist's+funding+as+MP+threatens+to+quit+post&pqatl=google). *Sunday Herald*. . Retrieved 29 December 2011. "Leaked correspondence from shadow Scottish secretary John Thurso MP, who favours nuclear power, suggests pressure is mounting within the party to reverse this policy."

External links

- John Thurso MP (http://www.johnthurso.org/) official site
- Profile (http://www.libdems.org.uk/people/john-thurso) at the Liberal Democrats
- Profile (http://www.scotlibdems.org.uk/people/mps/thurso) at the Scottish Liberal Democrats
- Hansard 1803–2005: contributions in Parliament by John Thurso (http://hansard.millbanksystems.com/people/viscount-thurso)
- Profile (http://www.parliament.uk/biographies/commons/john-thurso/25277) at Parliament of the United Kingdom
- Current session contributions in Parliament (http://www.publications.parliament.uk/pa/cm201213/cmhansrd/cmallfiles/mps/commons_hansard_4286_home.html) at *Hansard*

- Electoral history and profile (http://www.guardian.co.uk/politics/person/6594/john-thurso) at *The Guardian*
- Voting record (http://www.publicwhip.org.uk/mp.php?mpn=John_Thurso) at PublicWhip.org
- Record in Parliament (http://www.theyworkforyou.com/mp/john_thurso) at TheyWorkForYou.com
- Profile (http://www.parliamentaryrecord.com/content/profiles/mp/John-Thurso/Caithness-Sutherland-and-Easter-Ross/1127) at Westminster Parliamentary Record
- Profile (http://news.bbc.co.uk/democracylive/hi/representatives/profiles/25277.stm) at BBC News Democracy Live
- www.debretts.co.uk (http://www.debretts.co.uk/)

Douglas_Hogg

The Right Honourable **The Viscount Hailsham** PC QC		
[[file:DouglasHogg_20040917.jpg		alt=]]
Shadow Minister of Agriculture, Fisheries and Food		
In office 2 May 1997 – 17 June 1997		
Leader	John Major	
Preceded by	Jack Cunningham	
Succeeded by	David Curry	
Minister of Agriculture, Fisheries and Food		
In office 5 July 1995 – 2 May 1997		
Prime Minister	John Major	
Preceded by	William Waldegrave	
Succeeded by	Jack Cunningham	
Member of Parliament **for Sleaford and North Hykeham** **Grantham (1979–1997)**		
In office 4 May 1979 – 6 May 2010		
Preceded by	Joseph Godber	
Succeeded by	Stephen Phillips	
Personal details		
Born	5 February 1945Chelsea, London, England, United Kingdom	
Political party	Conservative	
Spouse(s)	Sarah Boyd-Carpenter (1968–present)	
Alma mater	Christ Church, Oxford	
Religion	Anglicanism	

Douglas Martin Hogg, 3rd Viscount Hailsham PC, QC (born 5 February 1945) is a British politician and barrister. A member of the Conservative Party, he served in the Cabinet as Minister of Agriculture, Fisheries and Food from 1995 to 1997, and was a Member of Parliament (MP) from 1979 to 2010. When the Daily Telegraph in 2009 exposed that Hogg had claimed upwards of £2,000 of taxpayers' money for the purposes of cleaning the moat around his country estate, Kettlethorpe Hall, he became one of the most prominent illustrations used by the media to portray the extent of the expenses scandal, though Hogg always maintained that the allegation was untrue. As a result of the negative publicity, Hogg did not seek re-election at the 2010 general election.

Early life

Douglas Hogg is the son of Lord Hailsham of St Marylebone, a former Lord Chancellor. He inherited the Viscountcy on 12 October 2001 upon the death of his father who had disclaimed that title for life in 1963, but who later accepted a life peerage in 1970; he is the grandson of Douglas Hogg, 1st Viscount Hailsham, also a former Lord Chancellor.

He was educated at Eton College and Christ Church, Oxford, where he graduated with a degree in History in 1968. In 1967, he served as the President of the Oxford Union. He was called to the Bar in 1968, after which he worked as a barrister. He became a Queen's Counsel in 1990, a year after his sister, Dame Mary Hogg, who is now a judge in the Family Division of the High Court.

Member of Parliament

He was elected as a Member of Parliament at the 1979 general election for the Lincolnshire seat of Grantham, following the retirement of the sitting Conservative MP Joseph Godber.

The Grantham seat was abolished at the 1997 general election; however, Hogg stood and was elected as MP for Sleaford and North Hykeham in 1997.

In government

In Parliament, Hogg served as a member of the Agriculture, Fisheries and Food Select Committee from 1979, until he was appointed as the Parliamentary Private Secretary (PPS) to the Chief Secretary to the Treasury Leon Brittan in 1982.

Hogg became a member of the government of Prime Minister Margaret Thatcher following the 1983 general election, when he was appointed as a Whip for a year. He rejoined the government in 1986 when he was appointed as a Parliamentary Under Secretary of State at the Home Office, and was promoted in 1989 to Minister of State at the Department of Trade and Industry.

Hogg was moved in 1990 under the leadership of Prime Minister John Major to the Foreign and Commonwealth Office, becoming a member of the Privy Council in 1992. He joined Major's Cabinet as the Minister of Agriculture, Fisheries and Food in 1995, serving in that capacity during the BSE crisis for which he received much criticism[1] and remaining in post until the election of the Tony Blair government in 1997.

On 3 March 1997, a disgruntled farmer from Anglesey, Louis Hayward, drove six hours from his farm to Hogg's house in Lincolnshire in order to dump three tonnes of pig manure outside his house.

Following the 1997 general election, Hogg was a member of the Home Affairs Select Committee for a year, and has remained on the backbenches since. The House of Lords Act 1999, which removed the right of hereditary peers to sit in the House of Lords, meant that Viscount Hailsham did not have to disclaim his peerage to remain an MP on the death of his father.

Stevens Inquiry

In the report of his inquiry concerning collusion in Northern Ireland between loyalist paramilitaries and the state security forces, under "Other Matters concerning Collusion", Sir John Stevens noted:

> "2.17 My Enquiry team also investigated an allegation that senior RUC officers briefed the Parliamentary Under-Secretary of State for the Home Department, the Rt Hon Douglas Hogg QC, MP, that 'some solicitors were unduly sympathetic to the cause of the IRA'. Mr Hogg repeated this view during a debate on the Prevention of Terrorism legislation in the House of Commons. Within a few weeks Patrick Finucane was murdered. Mr Hogg's comments about solicitors' support for terrorism made on 17 January 1989 aroused controversy. To the extent that they were based on information passed by the RUC, they were not justifiable and the Enquiry concludes that the Minister was compromised." [2]

Expenses

Even though he had been claiming near maximum Additional Costs Allowance in the 2001 and 2005 UK Parliaments, Douglas Hogg's total running costs as an MP (including allowances, expenses and staff costs) were consistently very much lower than most MPs.[3]

In 2009, during the row over MPs' expenses, *The Daily Telegraph* alleged that Hogg had submitted and was paid a claim form including more than £2,000 for the moat around his country estate, Kettleburgh Hall,[4] to be cleared.[5] The taxpayer helped meet the cost of a full-time housekeeper. Other allegations included expenses for work done to Hogg's stables and for his piano to be tuned.[6] It is claimed that the bills for Kettlethorpe Hall were so large, Hogg eventually agreed a deal with the expenses office simply to have one twelfth of the second homes allowance paid into his bank account every month.[4] In his defence, Hogg resolutely claimed he had not claimed for moat cleaning, and that the items were a list of all expenses incurred during house works, most of which were not paid for by the taxpayer. Hogg responded to the newspaper's claims by saying he had agreed the claims with the fees office, and therefore hoped and believed that they would comply with the rules and the 'spirit' of the rules.[7] In saying that his claims complied with both the spirit and letter of the rules,[8] Hogg said he had issued, in the interests of transparency, full lists of all his expenditure on the property but these were never meant to be the record of a claim.[9] On 14 May, Hogg agreed to repay the £2,200 cost of clearing the moat, after an order from the party leadership. He maintained he had not claimed the money but agreed it had not been "positively excluded" from paperwork submitted to the Commons fees office.[10]

Following the scandal, Hogg announced on 19 May 2009 that he would not stand at the next general election.[11] Prime Minister David Cameron put Hogg's name forward for a life barony to be included in the 2011 New Year Honours, but House of Lords Appointments Commission advised against the appointment.[12]

Personal life

He married Sarah, daughter of John Boyd-Carpenter, on 6 June 1968 in Westminster and they have a son (born October 1973) and a daughter (born August 1970). As his wife was created a life peeress, the Hailshams are one of the few couples to both hold noble titles in their own right.

References

[1] Richard Wilson (1998-12-16). "Minister without a friend" (http://news.bbc.co.uk/1/hi/uk_politics/236116.stm). *BBC News*. . Retrieved 2006-11-01.

[2] Stevens Inquiry 3 (http://www.madden-finucane.com/patfinucane/archive/pat_finucane/2003-04-17_stevens_report.pdf)

[3] "Douglas Hogg MP" (http://www.theyworkforyou.com/mp/douglas_hogg/sleaford_and_north_hykeham#expenses). *TheyWorkForYou*. mySociety is a project of UK Citizens Online Democracy (UKCOD). UKCOD is a registered charity in England and Wales, no. 1076346.. . Retrieved 2009-05-12.

[4] "http://www.mirror.co.uk/news/top-stories/2009/05/12/mps-expenses-abuse-all-stinks-115875-21351130" (http://www.mirror.co.uk/news/top-stories/2009/05/12/mps-expenses-abuse-all-stinks-115875-21351130/). Daily Mirror. 2009-05-14. . Retrieved 2009-05-14.

[5] Summers, Deborah (14 May 2009). "Andrew MacKay made 'unacceptable' home claims for eight or nine years" (http://www.guardian.co.uk/politics/2009/may/14/mps-expenses-hogg-morley). *The Guardian* (London). . Retrieved 1 May 2010.

[6] Winnett, Robert (2009-05-11). "MPs' expenses: Paying bills for Tory grandees" (http://www.telegraph.co.uk/news/newstopics/mps-expenses/5310200/MPs-expenses-Paying-bills-for-Tory-grandees.html). London: Daily Telegraph. . Retrieved 2009-05-11.

[7] Prince, Rosa (12 May 2009). "Daily Telegraph: Douglas Hogg (12 May 2009)" (http://www.telegraph.co.uk/news/newstopics/mps-expenses/5310069/MPs-expenses-Clearing-the-moat-at-Douglas-Hoggs-manor.html). London: Telegraph.co.uk. . Retrieved 2009-05-13.

[8] BBC News (http://news.bbc.co.uk/1/hi/uk_politics/8045384.stm)

[9] "Sleaford MP denies he claimed expenses for moat cleaning" (http://www.sleafordstandard.co.uk/news/Sleaford-MP-claimed-expenses-for.5256992.jp). *Sleaford Standard* (Sleaford, Lincolnshire). 2009-05-12. . Retrieved 2009-05-13. "MP Douglas Hogg has denied newspaper reports that he claimed more than £2,000 in expenses to have the moat at his Lincolnshire home cleaned.

...the Sleaford and North Hykeham MP told the Standard: "I have never claimed for the moat, or for the piano tuning - the allegation that I did is incorrect. I never claimed for these and I never received any money.

"The work to the stables that the Telegraph mentioned was actually for maintenance of security lights which were installed by the Home Office as part of the response to an IRA threat."

He said he had issued, in the interests of full transparency, full lists of all his expenditure on the property but these were never meant to be the record of a claim.

Mr Hogg said it was clear to the fees office that the overall allowable expenses were over the Additional Costs Allowance and that his claim only covered utilities, council tax, building insurance, the alarm system, heating, repairs and maintenance of house and garden and 65 per cent of the cost of a housekeeper to clean and maintain the house and look after it when he and his wife, Baroness Hogg, were away.

He said: "It was on this basis and with the express agreement of the fees office, in advance and in writing that I was making a monthly claim equal to one-twelfth of the ACA."

Mr Hogg added: "I am amongst the lowest claimers in parliament – specifically 551st out of 645 MPs in 2007/8.

"There is no doubt that our system has lost public confidence and we as parliamentarians have got to accept that we are responsible for having put the system in place and that it is probably flawed.

"We got it wrong and need to apologise for that, and I do apologise for it.""

[10] "Tory MP Hogg to repay 'moat cost'" (http://news.bbc.co.uk/1/hi/uk_politics/8051027.stm). BBC News. 2009-05-14. . Retrieved 2009-05-14.

[11] "Moat claim MP to quit at election" (http://news.bbc.co.uk/2/hi/uk_news/politics/8057579.stm). BBC News Online. 2009-05-19. . Retrieved 2009-05-19.

[12] Gammell, Caroline (6 March 2011). "Tory MP who claimed for moat cleaning is denied peerage" (http://www.telegraph.co.uk/news/8364609/Tory-MP-who-claimed-for-moat-cleaning-is-denied-peerage.html). *The Daily Telegraph* (London). .

External links

- Conservative Party - Rt Hon Douglas Hogg QC MP (http://www.conservatives.com/tile.do?def=people.person.page&PersonID=4642) biography
- Guardian Unlimited Politics - Ask Aristotle: Douglas Hogg MP (http://politics.guardian.co.uk/person/0,9290,-2461,00.html)
- TheyWorkForYou.com - Douglas Hogg MP (http://www.theyworkforyou.com/mp/douglas_hogg/sleaford_and_north_hykeham)
- Hansard 1803–2005: contributions in Parliament by Douglas Hogg (http://hansard.millbanksystems.com/people/mr-douglas-hogg)
- The Public Whip - Douglas Hogg MP (http://www.publicwhip.org.uk/mp.php?mpn=Douglas_Hogg&mpc=Sleaford+&+North+Hykeham) voting record
- BBC News - Douglas Hogg (http://news.bbc.co.uk/1/shared/mpdb/html/404.stm) profile 15 February 2005
- Portraits of Douglas Martin Hogg (http://www.npg.org.uk/collections/search/person.php?LinkID=mp99240) at the National Portrait Gallery, London

Viscount_Cobham

Viscount Cobham is a title in the Peerage of Great Britain. It was created in 1718 for Field Marshal Sir Richard Temple, 1st Baron Cobham and 4th Baronet, of Stowe. The Viscount holds the subsidiary titles of **Baron Cobham**, of Cobham in the County of Kent, (1718), **Baron Westcote**, of Ballymore in the County of Longford (1776), and **Baron Lyttelton**, of Frankley in the County of Worcester (1794). All the titles are in the Peerage of Great Britain, except the barony of Westcote, which is in the Peerage of Ireland. The Viscount is also a **Baronet**, of Frankley in the County of Worcester (1618).

The barony and viscountcy of Cobham were subsidiary titles of the earldom of **Temple** from 1750 to 1784, of the **Marquessate of Buckingham** from 1784 to 1822 and of the **Dukedom of Buckingham and Chandos** from 1822 to 1889. Since the latter year the titles are merged with the barony of Lyttelton and barony of Westcote.

History

Richard Temple, 1st Viscount Cobham

The Temple family descended from **Peter Temple** of Burton Dassett. His younger son Anthony Temple founded the Irish branch of the family from whom the Viscounts Palmerston descended.[1] Peter Temple's eldest son, **John Temple**, acquired the Stowe estate in Buckinghamshire. The latter's son Thomas Temple represented Andover in Parliament. In 1611 he was created a **Baronet**, of Stowe in the County of Buckingham, in the Baronetage of England. His son, the second Baronet, represented Buckingham in both the Short Parliament and the Long Parliament. He was succeeded by his son, the third Baronet. He sat in Parliament for Warwickshire and Buckingham.

His son, the fourth Baronet *(pictured)*, was a noted soldier and politician. In 1714 he was raised to the Peerage of Great Britain as **Baron Cobham**, of Cobham in the County of Kent, with remainder to heirs male of his body. Four years later he was created **Baron Cobham**, of Cobham in the County of Kent, and **Viscount Cobham**, with remainder, failing heirs male of his own, to (1) his sister Hester Temple and the heirs male of her body and failing which to (2) his third sister Lady Christian, wife of Sir Thomas Lyttelton, 4th Baronet, of Frankley (see the Baron Lyttelton). Lord Cobham died childless and was succeeded in the baronetcy by his cousin, the fifth Baronet, the great-grandson of Sir John Temple, second son of the first Baronet. The baronetcy became dormant in 1786 on the death of the seventh Baronet. Furthermore, on Lord Cobham's death the barony of 1714 became extinct while he was succeeded in the barony and viscountcy of 1718 according to the special remainder by his sister Hester Temple. She was the widow of Richard Grenville. In 1751 she was created **Countess Temple** in the Peerage of Great Britain, with remainder to the heirs male of her body. Lady Temple's younger son was Prime Minister George Grenville.

She was succeeded by her eldest son, the second Earl. He inherited the Temple estates, including Stowe House, and assumed the additional surname of Temple. Lord Temple was also involved in politics and held office as First Lord of the Admiralty and as Lord Privy Seal. On his death the titles passed to his nephew, the third Earl, the son of George Grenville. He served as Lord Lieutenant of Ireland between 1782 and 1783 and 1787 and 1789. In 1784 he was created **Marquess of Buckingham** in the Peerage of Great Britain. Lord Buckingham married Lady Mary Nugent, daughter of Robert Nugent, 1st Earl Nugent. Mary was in 1800 created Baroness Nugent in her own right in the Peerage of Ireland, with remainder to her second son George (see the Baron Nugent). In 1788 Lord Buckingham also succeeded his father-in-law as second Earl Nugent according to a special remainder in the letters patent. He assumed by Royal license the additional surname of Nugent at the same time.

He was succeeded by his son, the second Marquess. He was also involved in politics and served as Joint Paymaster of the Forces from 1806 to 1807. He married Lady Anne Eliza Brydges, the only child of James Brydges, 3rd Duke of Chandos (a title which became extinct on his death in 1789), and assumed by Royal license the additional surnames of Brydges-Chandos in 1799. In 1822 Lord Buckingham was created **Earl Temple of Stowe**, in the County of Buckingham, **Marquess of Chandos** and **Duke of Buckingham and Chandos**, all in the Peerage of the United Kingdom. The earldom was created with remainder, failing male issue of his own, to (1) the heirs male of the body of his deceased great-grandmother Hester Grenville, 1st Countess Temple, and (2) in default thereof to his granddaughter Lady Anne Eliza Mary Temple-Nugent-Brydges-Chandos-Grenville, daughter of his son Richard, Earl Temple, who succeeded as second Duke. He was a Tory politician and served as Lord Privy Seal from 1841 to 1842.

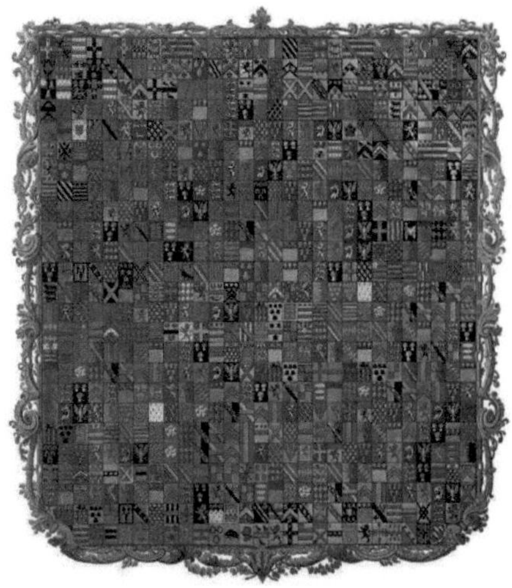

The 719 heraldic quarterings of the Temple-Nugent-Brydges-Chandos-Grenville family.

On his death the titles passed to his son, the third Duke. He was also a prominent politician and served as Lord President of the Council and as Secretary of State for the Colonies. In 1868 the Duke established his right to the Scottish lordship of Kinloss before the Committee for Privileges of the House of Lords. However, on his death in 1889 without male issue, the dukedom and its subsidiary titles the marquessate of Buckingham, marquessate of Chandos, earldom of Temple and earldom of Nugent became extinct. The earldom of Temple of Stowe and lordship of Kinloss also held by the Duke were passed on to his nephew and daughter respectively (see these titles for more information).

The barony and viscountcy of Cobham also survived, and were inherited according to the special remainders by the Duke's distant relative Charles Lyttelton, 5th Baron Lyttelton, who became the eighth Viscount Cobham. He was the great-great-grandson of the aforementioned Lady Christian and Sir Thomas Lyttelton, 4th Baronet, of Frankley (see the Baron Lyttelton for earlier history of the Lyttelton family). The eighth Viscount had previously represented East Worcestershire in Parliament as a Liberal. His son, the ninth Viscount, was Lord Lieutenant of Worcestershire from 1923 to 1949. He was succeeded by his son, the tenth Viscount. He notably served as Governor-General of New Zealand from 1957 to 1962. As of 2010 the titles are held by his younger son, the twelfth Viscount, who succeeded his elder brother in 2006.

The family seat is Hagley Hall, near Stourbridge in Worcestershire.

Another member of the Grenville family was Prime Minister William Wyndham Grenville, 1st Baron Grenville. He was the younger son of George Grenville and the younger brother of the first Marquess of Buckingham. Another member of the Lyttelton family was the Hon. Alfred Lyttelton. He was the eighth son of the fourth Baron Lyttelton. His son was the politician Oliver Lyttelton, 1st Viscount Chandos. Yet another member of the family was the jazz musician Humphrey Lyttelton. He was the son of the Hon. George William Lyttelton, second son of the eighth Viscount Cobham.

Temple Baronets, of Stowe (1611)

- Sir Thomas Temple, 1st Baronet (1567–1637)
- Sir Peter Temple, 2nd Baronet (1592–1653)
- Sir Richard Temple, 3rd Baronet (1634–1697)
- Sir Richard Temple, 4th Baronet (1669–1749) (created **Viscount Cobham** in 1718; see below)
- Sir William Temple, 5th Baronet (1694–1760)
- Sir Peter Temple, 6th Baronet (d. 1761)
- Sir Richard Temple, 7th Baronet (1731–1786) (dormant)

Viscounts Cobham (1718)

- Richard Temple, 1st Viscount Cobham (1669–1749)
- Hester Grenville, 2nd Viscountess Cobham (c. 1690–1752) (created **Countess Temple** in 1751)

Earls Temple (1751)

- Hester Grenville, 1st Countess Temple, 2nd Viscountess Cobham (c. 1690–1752)
- Richard Grenville-Temple, 2nd Earl Temple, 3rd Viscount Cobham (1711–1779)
- George Nugent-Temple-Grenville, 3rd Earl Temple, 4th Viscount Cobham (1753–1813) (created **Marquess of Buckingham** in 1784)

Marquesses of Buckingham (1784)

- George Nugent-Temple-Grenville, 1st Marquess of Buckingham, 4th Viscount Cobham (1753–1813)
- Richard Temple-Nugent-Brydges-Chandos-Grenville, 2nd Marquess of Buckingham, 5th Viscount Cobham (1776–1839) (created **Duke of Buckingham and Chandos** in 1822)

Dukes of Buckingham and Chandos (1822)

- Richard Temple-Nugent-Brydges-Chandos-Grenville, 1st Duke of Buckingham and Chandos, 5th Viscount Cobham (1776–1839)
- Richard Plantagenet Temple-Nugent-Brydges-Chandos-Grenville, 2nd Duke of Buckingham and Chandos, 6th Viscount Cobham (1797–1861)
- Richard Plantagenet Campbell Temple-Nugent-Brydges-Chandos-Grenville, 3rd Duke of Buckingham and Chandos, 7th Viscount Cobham (1823–1889)

Viscounts Cobham (1718; Reverted)

- Charles George Lyttelton, 8th Viscount Cobham (1842–1922)
- John Cavendish Lyttelton, 9th Viscount Cobham (1881–1949)
- Charles John Lyttelton, 10th Viscount Cobham (1909–1977)
- John William Leonard Lyttelton, 11th Viscount Cobham (1943–2006)
- Christopher Charles Lyttelton, 12th Viscount Cobham (b. 1947)

The heir apparent is the present holder's son Hon. Oliver Christopher Lyttelton (b. 1976)

See also

- Duke of Chandos
- Earl Temple of Stowe
- Lady Kinloss
- Baron Grenville
- Earl Nugent
- Viscount Palmerston
- Baron Lyttelton
- Viscount Chandos
- Temple Baronets
- Baron Cobham

External links

- A website about the Dukes of Buckingham and Chandos [2]

References

- Kidd, Charles, Williamson, David (editors). *Debrett's Peerage and Baronetage* (1990 edition). New York: St Martin's Press, 1990.
- Leigh Rayment's Peerage Pages [1]

[1] www.stirnet.com/HTML/genie/british/tt/temple01.htm
[2] http://www.dukesofbuckingham.org.uk/

Viscount_Chandos

Viscount Chandos, of Aldershot in the County of Southampton, is a title in the Peerage of the United Kingdom. It was created in 1954 for the businessman and public servant Oliver Lyttelton. He was the son of the politician and sportsman the Honourable Alfred Lyttelton, eighth son of George Lyttelton, 4th Baron Lyttelton, whose eldest son, the fifth Baron, succeeded his kinsman Richard Temple-Nugent-Brydges-Chandos-Grenville, 3rd Duke of Buckingham and Chandos as 8th Viscount Cobham in 1889. As of 2010 the title is held by the first Viscount's grandson, the third Viscount, who succeeded his father in 1980. He lost his seat in the House of Lords after the passing of the House of Lords Act of 1999, which removed the automatic right of hereditary peers to sit in the upper chamber of Parliament. However, in 2000 he was given a life peerage as **Baron Lyttelton of Aldershot**, of Aldershot in the County of Hampshire, and was able to return to the House of Lords, where he now sits on the Labour benches. Lord Chandos is also in remainder to the viscountcy of Cobham and its subsidiary titles the barony of Cobham, the barony of Lyttelton, the barony of Westcote and the baronetcy of Frankley.

Viscounts Chandos (1954)

- Oliver Lyttelton, 1st Viscount Chandos (1893–1972)
- Anthony Lyttelton, 2nd Viscount Chandos (1920–1980)
- Thomas Orlando Lyttelton, 3rd Viscount Chandos (b. 1953)

The heir apparent is the present holder's son Oliver Antony Lyttelton (b. 1986)

See also

- Viscount Cobham
- Baron Lyttelton
- Duke of Buckingham and Chandos
- Duke of Chandos

Notes

- Kidd, Charles, Williamson, David (editors). *Debrett's Peerage and Baronetage* (1990 edition). New York: St Martin's Press, 1990.
- Leigh Rayment's Peerage Pages [1]

External links

- Hansard 1803–2005: contributions in Parliament by Anthony Lyttelton, 2nd Viscount Chandos [1]

References

[1] http://hansard.millbanksystems.com/people/mr-antony-lyttelton

Liberal_Party_(UK)

Liberal Party	
Founded	1859
Dissolved	2 March 1988
Preceded by	Whigs, Radicals, Peelites
Merged into	Liberal Democrats
Ideology	*Historical:* Classical liberalism *Modern:* Social liberalism
International affiliation	Liberal International
Colours	Yellow; historically orange, and before that green

The **Liberal Party** was one of the two major political parties of the United Kingdom during the 19th and early 20th centuries. It was the third largest political party throughout the latter half of the 20th century. The party moved towards social liberalism during the late 19th and early 20th centuries, and came to be seen as being on the centre-left of British politics.

It grew out of the Whig Party in 1859, and William Gladstone carried the party through governments in the late 19th century. The party split over the issue of Irish Home Rule, but returned to power in 1906 through a landslide victory. During this time in government, the Liberals are credited with the so-called Liberal Reforms, which saw the creation of a basic welfare state. During World War I, the Liberals governed Britain through a coalition with the Conservatives, which ended in 1922. In the 1920s, the Labour Party permanently replaced the Liberals as the largest opponent of the Conservative Party in British politics, and the Liberals went into decline, which culminated in their winning as few as 6 seats at general elections during the 1950s. A comeback of sorts was seen during the 1960s and 1970s, but it was not until 1981, when the Liberals allied with the newly formed SDP to form the SDP-Liberal Alliance, that the party enjoyed significant electoral success. In the 1983 general election, the alliance took 25% of the popular vote, but this only translated into 23 seats (out of a total of 650). During 1988, the Liberals formally merged with the SDP to form the Liberal Democrats, though a small splinter Liberal Party was formed in 1989 by former members opposed to the merger.

Ideology

During the 19th century the Liberal Party was broadly in favour of what would today be called classical liberalism: supporting *laissez-faire* economic policies such as free trade and minimal government interference in the economy (this doctrine was usually termed 'Gladstonian Liberalism' after the Victorian era Liberal Prime Minister William Ewart Gladstone). The Liberal Party favoured social reform, personal liberty, reducing the powers of the Crown and the Church of England (many of them were Nonconformists) and an extension of the electoral franchise. Sir William Harcourt, a prominent Liberal politician in the Victorian era, said this about liberalism in 1872:

A crowd waits outside Leeds Town Hall to see them elect a Liberal Party candidate during the 1880 general elections.

> If there be any party which is more pledged than another to resist a policy of restrictive legislation, having for its object social coercion, that party is the Liberal party. (Cheers.) But liberty does not consist in making others do what you think right, (Hear, hear.) The difference between a free Government and a Government which is not free is principally this—that a Government which is not free interferes with everything it can, and a free Government interferes with nothing except what it must. A despotic Government tries to make everybody do what it wishes; a Liberal Government tries, as far as the safety of society will permit, to allow everybody to do as he wishes. It has been the tradition of the Liberal party consistently to maintain the doctrine of individual liberty. It is because they have done so that England is the place where people can do more what they please than in any other country in the world...It is this practice of allowing one set of people to dictate to another set of people what they shall do, what they shall think, what they shall drink, when they shall go to bed, what they shall buy, and where they shall buy it, what wages they shall get and how they shall spend them, against which the Liberal party have always protested.[1]

The political terms of "modern", "progressive" or "new" Liberalism began to appear in the mid to late 1880s and became increasingly common to denote the tendency in the Liberal Party to favour an increased role for the state as more important than the classical liberal stress on self-help and freedom of choice.[2]

By the early 20th century the Liberals stance began to shift towards "New Liberalism", what would today be called social liberalism: a belief in personal liberty with a support for government intervention to provide minimum levels of welfare.[3] This shift was best exemplified by the Liberal government of Herbert Henry Asquith and his Chancellor David Lloyd George, whose Liberal reforms in the early 1900s created a basic welfare state.[4]

David Lloyd George adopted a programme at the 1929 general election entitled *We Can Conquer Unemployment!*, although by this stage the Liberals had declined to third-party status. The Liberals now (as expressed in the Liberal Yellow Book) regarded opposition to state intervention as being a characteristic of right-wing extremists.[5]

After nearly becoming extinct in the 1940s and 50s, the Liberal Party revived its fortunes somewhat under the leadership of Jo Grimond in the 1960s, by positioning itself as a radical centrist non-socialist alternative to the Conservative and Labour Party governments of the time.[6]

Origins

Viscount Palmerston

The Liberal Party grew out of the Whigs, which had its origins as an aristocratic faction in the reign of Charles II. The Whigs were in favour of reducing the power of The Crown and increasing the power of the Parliament. Although their motives in this were originally to gain more power for themselves, the more idealistic Whigs gradually came to support an expansion of democracy for its own sake. The great figures of reformist Whiggery were Charles James Fox (died 1806) and his disciple and successor Earl Grey. After decades in opposition the Whigs came to power under Grey in 1830 and carried the First Reform Act in 1832.

The Reform Act was the climax of Whiggery but also brought about the Whigs' demise. The admission of the middle classes to the franchise and to the House of Commons led eventually to the development of a systematic middle class liberalism and the end of Whiggery, although for many years reforming aristocrats held senior positions in the party. In the years after Grey's retirement the party was led first by Lord Melbourne, a fairly traditional Whig, and then by Lord John Russell, the son of a Duke but a crusading radical, and Lord Palmerston, a renegade Irish Tory and essentially a conservative, although capable of radical gestures.

As early as 1839 Russell had adopted the name Liberal Party, but in reality the party was a loose coalition of Whigs in the House of Lords and Radicals in the Commons. The leading Radicals were John Bright and Richard Cobden, who represented the manufacturing towns that had gained representation under the Reform Act. They favoured social reform, personal liberty, reducing the powers of the Crown and the Church of England (many of them were Nonconformists), avoidance of war and foreign alliances (which were bad for business), and above all free trade. For a century free trade was the one cause which could unite all Liberals.

In 1841 the Liberals lost office to the Conservative Party under Sir Robert Peel, but their period in opposition was short, because the Conservatives split over the repeal of the Corn Laws, a free trade issue, and a faction known as the Peelites (but not Peel himself, who died soon after), defected to the Liberal side. This allowed ministries led by Russell, Palmerston and the Peelite Lord Aberdeen to hold office for most of the 1850s and 1860s. The leading Peelite was William Ewart Gladstone, who was a reforming Chancellor of the Exchequer in most of these governments. The formal foundation of the Liberal Party is traditionally traced to 1859 and the formation of Palmerston's second government.

The Whig-Radical amalgam could not become a true modern political party, however, while it was dominated by aristocrats, and it was not until the departure of the "Two Terrible Old Men", Russell and Palmerston, that Gladstone could become the first leader of the modern Liberal Party. This was brought about by Palmerston's death in 1865 and Russell's retirement in 1868. After a brief Conservative interlude (during which the Second Reform Act was passed by agreement between the parties), Gladstone won a huge victory at the 1868 election and formed the first Liberal government. The establishment of the party as a national membership organisation came with the foundation of the National Liberal Federation in 1877.

The Gladstonian era

William Gladstone

For the next thirty years Gladstone and Liberalism were synonymous. William E. Gladstone served as prime minister four times (1868–74, 1880–85, 1886, and 1892–94). His financial policies, based on the notion of balanced budgets, low taxes and laissez-faire, were suited to a developing capitalist society but could not respond effectively as economic and social conditions changed. Called the "Grand Old Man" later in life, he was always a dynamic popular orator who appealed strongly to British workers and lower middle class. The deeply religious Gladstone brought a new moral tone to politics with his evangelical sensibility and opposition to aristocracy.[7] His moralism often angered his upper-class opponents (including Queen Victoria), and his heavy-handed control split the Liberal party.[8] [9]

In foreign policy, Gladstone was in general against foreign entanglements but did not resist the reality of imperialism. For example, he approved of the occupation of Egypt by British forces in 1882. His goal was to create a European order based on cooperation rather than conflict and mutual trust instead of rivalry and suspicion; the rule of law was to supplant the reign of force and self-interest. This Gladstonian concept of a harmonious Concert of Europe was opposed to and ultimately defeated by a Bismarckian system of manipulated alliances and antagonisms.[10]

As prime minister 1868 to 1874 he headed a Liberal Party that was a coalition of Peelites like himself, Whigs and radicals; Gladstone was now a spokesman for "peace, economy and reform." One major achievement was the Elementary Education Act of 1870. It provided England with an adequate system of elementary schools for the first time and required attendance. He also secured the abolition of the purchase of commissions in the army, and of religious tests for admission to Oxford and Cambridge; the introduction of the secret ballot in elections; the legalization of trade unions; and the reorganization of the judiciary in the Judicature Act.[11]

Regarding Ireland, the major Liberal achievements were land reform, where he ended centuries of landlord oppression, and the disestablishment of the (Anglican) Church of Ireland through the Irish Church Act 1869.

In the 1874 general election Gladstone was defeated by the Conservatives under Disraeli during a sharp economic recession. He formally resigned as Liberal leader and was succeeded by the Marquess of Hartington, but he soon changed his mind and returned to active politics. He strongly disagreed with Disraeli's pro-Ottoman foreign policy and in 1880 he conducted the first outdoor mass-election campaign in Britain, known as the Midlothian campaign. The Liberals won a large majority in the 1880 election. Hartington ceded his place and Gladstone resumed office.

Among the consequences of the Third Reform Act (1884–85) was the giving of the vote to the Catholic peasants in Ireland, and the consequent creation of an Irish Parliamentary Party led by Charles Stewart Parnell. In the 1885 general election this party won the balance of power in the House of Commons, and demanded Irish Home Rule as the price of support for a continued Gladstone ministry. Gladstone personally supported Home Rule, but a strong Liberal Unionist faction led by Joseph Chamberlain, along with the last of the Whigs, Hartington, opposed it. The Irish Home Rule bill gave all owners of Irish land a chance to sell to the state at a price equal to 20 years' purchase of the rents and allowing tenants to purchase the land. Irish nationalist reaction was mixed, Unionist opinion was hostile, and the election addresses during the 1886 election revealed English radicals to be against the bill also. Among the Liberal rank and file, several Gladstonian candidates disowned the bill, reflecting fears at the constituency level that the interests of the working people were being sacrificed to finance a rescue operation for the landed elite.[12]

The result was a catastrophic split in the Liberal Party, and heavy defeat in the 1886 election at the hands of Lord Salisbury. There was a final weak Gladstone ministry in 1892, but it also was dependent on Irish support and failed to get Irish Home Rule through the House of Lords.

Gladstone finally retired in 1894, and his ineffectual successor, Lord Rosebery, led the party to another heavy defeat in the 1895 general election.

A major long-term consequence of the Third Reform Act was the rise of Lib-Lab candidates, in the absence of any committed Labour Party. The Act split all county constituencies (which were represented by multiple MPs) in to single-member constituencies, roughly corresponding to population patterns. In areas with working class majorities, in particular coal-mining areas, Lib-Lab candidates were popular, and they received sponsorship and endorsement from trade unions. In the first election after the Act was passed (1885), thirteen were elected, up from two in 1874. The Third Reform Act also facilitated the demise of the Whig old guard: in two-member constituencies, it was common to pair a Whig and a radical under the Liberal banner. After the Third Reform Act, fewer Whigs were selected as candidates.[13]

The Liberal zenith

Herbert Henry Asquith

The Liberals languished in opposition for a decade, while the coalition of Salisbury and Chamberlain held power. The 1890s were marred by infighting between the three principal successors to Gladstone, party leader William Harcourt, former Prime Minister Lord Rosebery, and Gladstone's personal secretary, John Morley. This intrigue finally led Harcourt and Morley to resign their positions in 1898 as they continued to be at loggerheads with Rosebery over Irish home rule and issues relating to imperialism. Replacing Harcourt as party leader was Sir Henry Campbell-Bannerman. Harcourt's resignation briefly muted the turmoil in the party, but the beginning of the Second Boer War soon nearly broke the party apart, with Rosebery and a circle of supporters including important future Liberal leaders H.H. Asquith, Edward Grey, and Richard Burdon Haldane forming a clique dubbed the "Liberal Imperialists" that supported the government in the prosecution of the war. On the other side, more radical members of the party formed a Pro-Boer faction that denounced the conflict and called for an immediate end to hostilities. Quickly rising to prominence among the Pro-Boers was David Lloyd George, a relatively new MP and a master of rhetoric, who took advantage of having a national stage to speak out on a controversial issue to make his name in the party. Harcourt and Morley also sided with this group, though with slightly different aims. Campbell-Bannerman tried to keep these forces together at the head of a moderate Liberal rump, but in 1901 he delivered a speech on the government's "methods of barbarism" in South Africa that pulled him further to the left and nearly tore the party in two. The party was saved after Salisbury's retirement in 1902 when his successor, Arthur Balfour, pushed a series of unpopular initiatives such as a new education bill and Joseph Chamberlain called for a new system of protectionist tariffs. Campbell-Bannerman was able to rally the party around the traditional liberal platform of free trade and land reform and led them to the greatest election victory in their history. This would prove the last time the Liberals won a majority in their own right.

Although he presided over a large majority, Sir Henry Campbell-Bannerman was overshadowed by his ministers, most notably Herbert Henry Asquith at the Exchequer, Edward Grey at the Foreign Office, Richard Burdon Haldane at the War Office and David Lloyd George at the Board of Trade. An ill Campbell-Bannerman retired in 1908 and died later that year. He was succeeded by Asquith, who stepped up the government's radicalism. Lloyd George succeeded Asquith at the Exchequer, and was in turn succeeded at the Board of Trade by Winston Churchill, a recent defector from the Conservatives.

The Liberals pushed through much legislation, including the regulation of working hours, National Insurance and welfare. It was at this time that a political battle over the so-called People's Budget resulted in the passage of an act ending the power of the House of Lords to block legislation. The cost was high, however, as the government was required by the king to call two general elections in 1910 to validate its position and ended up frittering away most of its large majority, being left once again dependent on the Irish Nationalists.

As a result Asquith was forced to introduce a new third Home Rule bill in 1912. Since the House of Lords no longer had the power to block the bill, the Unionist's Ulster Volunteers led by Sir Edward Carson, launched a campaign of opposition that included the threat of armed resistance in Ulster and the threat of mutiny by army officers in Ireland in 1914 (*see Curragh Incident*). In their resistance to Home Rule the Ulster Protestants had the full support of the Conservatives, whose leader, Andrew Bonar Law, was of Ulster-Scots descent. The country seemed to be on the brink of civil war when World War I broke out in August 1914.

The war struck at the heart of everything British Liberals believed in. The party became divided over the distinctly illiberal policies that were introduced under her auspices, including conscription and the Defence of the Realm Act. Several Cabinet ministers resigned, and Asquith, the master of domestic politics, proved a poor war leader. Lloyd George and Churchill, however, were zealous supporters of the war, and gradually forced the old pacifist Liberals out. The poor British performance in the early months of the war forced Asquith to invite the Conservatives into a coalition (on 17 May 1915). This marked the end of the last all-Liberal government. This coalition fell apart at the end of 1916, when the Conservatives withdrew their support from Asquith and gave it to Lloyd George instead, who became Prime Minister at the head of a coalition government largely made up of Conservatives. Asquith and his followers moved to the opposition benches in Parliament and the Liberal Party was split once again.

Liberal decline

David Lloyd George

In the 1918 general election Lloyd George, "the Man Who Won the War", led his coalition into another *khaki election*, and won a sweeping victory over the Asquithian Liberals and the newly emerging Labour Party. Lloyd George and the Conservative leader Andrew Bonar Law wrote a joint letter of support to candidates to indicate they were considered the official Coalition candidates – this "coupon" as it became known was issued against many sitting Liberal MPs, often to devastating effect, though not against Asquith himself. Asquith and most of his colleagues lost their seats. Lloyd George still claimed to be leading a Liberal government, but he was increasingly under the influence of the rejuvenated Conservative party. In 1922 the Conservative backbenchers rebelled against the continuation of the coalition, citing in particular the Chanak Crisis over Turkey and Lloyd George's corrupt sale of honours amongst other grievances, and Lloyd George was forced to resign. The Conservatives came back to power under Bonar Law and then Stanley Baldwin.

At the 1922 and 1923 elections the Liberals won barely a third of the vote and only a quarter of the seats in the House of Commons, as many radical voters abandoned the divided Liberals and went over to Labour. In 1922 Labour became the official opposition. A reunion of the two warring factions took place in 1923 when the new Conservative Prime Minister Stanley Baldwin committed his party to protective tariffs, causing the Liberals to reunite in support of free trade. The party gained ground in the 1923 general election but ominously made most of its gains from Conservatives whilst losing ground to Labour – a sign of the party's direction for many years to come. The party remained the third largest in the House of Commons, but the Conservatives had lost their majority. There was much speculation and fear about the prospect of a Labour government, and comparatively little about a Liberal government, even though it could have plausibly presented an experienced team of ministers compared to Labour's almost complete lack of experience, as well as offering a middle ground that could get support from both Conservatives and Labour in crucial Commons divisions. But instead of trying to force the opportunity to form a Liberal government, Asquith decided instead to allow Labour the chance of office in the belief that they would prove incompetent and this would set the stage for a revival of Liberal fortunes at Labour's expense. It was a fatal error.

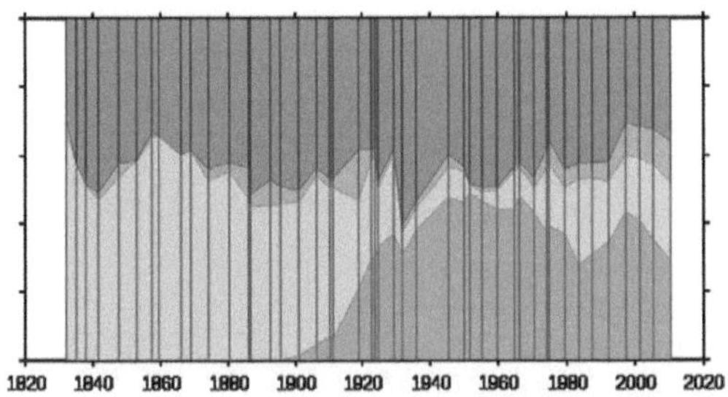

Share of the vote received by Conservatives (blue), Whigs/Liberals/Liberal Democrats (orange), Labour (red) and others (grey) in general elections since 1832.[14] [15] Following success as the successor to the Whig party, the party's share of the popular vote plummeted after the First World War as it lost votes to the new Labour party and fractured into groups such as the National and Coalition Liberals.

Labour was determined to destroy the Liberals and become the sole party of the left. Ramsay MacDonald was forced into a snap election in 1924, and although his government was defeated, he achieved his objective of virtually wiping the Liberals out as many more radical voters now moved to Labour whilst moderate middle-class Liberal voters concerned about socialism moved to the Conservatives. The Liberals were reduced to a mere forty seats in Parliament, only seven of which had been won against candidates from both parties and none of these formed a coherent area of Liberal survival. The party seemed finished and during this period some Liberals, such as Churchill, went over to the Conservatives, while others went over to Labour. (Several Labour ministers of later generations, such as Michael Foot and Tony Benn, were the sons of Liberal MPs.)

Asquith died in 1928 and the enigmatic figure of Lloyd George returned to the leadership and began a drive to produce coherent policies on many key issues of the day. In the 1929 general election he made a final bid to return the Liberals to the political mainstream, with an ambitious programme of state stimulation of the economy called *We Can Conquer Unemployment!*, largely written for him by the Liberal economist John Maynard Keynes. The Liberals gained ground, but once again it was at the Conservatives' expense whilst also losing seats to Labour. Indeed the urban areas of the country suffering heavily from unemployment, which might have been expected to respond the most to the radical economic policies of the Liberals, instead gave the party its worst results. By contrast most of the party's seats were won either due to the absence of a candidate from one of the other parties or in rural areas on the "Celtic fringe", where local evidence suggests that economic ideas were at best peripheral to the electorate's concerns. The Liberals now found themselves with 59 members holding the balance of power in a Parliament where Labour was the largest party but lacked an overall majority. Lloyd George offered a degree of support to the Labour government in the hope of winning concessions, including a degree of electoral reform to introduce the alternative vote, but this support was to prove bitterly divisive as the Liberals increasingly divided between those seeking to gain what Liberal goals they could achieve, those who preferred a Conservative government to a Labour one and vice-versa.

The last majority Liberal Government in Britain was elected in 1906. The pre-World War I period in Britain was a period of worker strikes and civil unrest involving many violent confrontations between the police/troops, and civilians. Other "issues of the period" included the suffragettes and the Irish Independence movement. After the carnage of 1914–1918 the democratic reform of the Representation of the People Act 1918 instantly tripled the number of people allowed to vote in Britain from 7 to 21 million. The Labour Party benefited the most from this huge change in the British electorate, forming its first minority government in 1924.

The splits over the National Government

Herbert Samuel

In 1931 MacDonald's government fell apart under the Great Depression, and the Liberals agreed to join his National Government, dominated by the Conservatives. Lloyd George himself was ill and did not actually join. Soon, however, the Liberals faced another divisive crisis when a National Government was proposed to fight the 1931 general election with a mandate for tariffs. From the outside, Lloyd George called for the party to abandon the government completely in defence of free trade, but only a few MPs and candidates followed. Another group under Sir John Simon then emerged, who were prepared to continue their support for the government and take the Liberal places in the Cabinet if there were resignations. The third group under Sir Herbert Samuel pressed for the parties in government to fight the election on separate platforms. In doing so the bulk of Liberals remained supporting the government, but two distinct Liberal groups had emerged within this bulk – the Liberal Nationals (officially the "National Liberals" after 1947) led by Simon, also known as "Simonites", and the "Samuelites" or "official Liberals", led by Samuel who remained as the official party. Both groups secured about 34 MPs but proceeded to diverge even further after the election, with the Liberal Nationals remaining supporters of the government throughout its life. There were to be a succession of discussions about them rejoining the Liberals, but these usually foundered on the issues of free trade and continued support for the National Government. The one significant reunification came in 1946 when the Liberal and Liberal National party organisations in London merged.

The official Liberals found themselves a tiny minority within a government committed to protectionism. Slowly they found this issue to be one they could not support. In early 1932 it was agreed to suspend the principle of collective responsibility to allow the Liberals to oppose the introduction of tariffs. Later in 1932 the Liberals resigned their ministerial posts over the introduction of the Ottawa Agreement on Imperial Preference. However, they remained sitting on the government benches supporting it in Parliament, though in the country local Liberal activists bitterly opposed the government. Finally in late 1933 the Liberals crossed the floor of the House of Commons and went into complete opposition. By this point their number of MPs was severely depleted. In the 1935 general election, just 17 Liberal MPs were elected, along with Lloyd George and three followers as "independent Liberals". Immediately after the election the two groups reunited, though Lloyd George declined to play much of a formal role in his old party. Over the next ten years there would be further defections as MPs deserted to either the Liberal Nationals or Labour. Yet there were a few recruits, such as Clement Davies, who had deserted to the National Liberals in 1931 but now returned to the party during the World War II and who would lead it after the war.

Near extinction

Samuel had lost his seat in the 1935 election and the leadership of the party fell to Sir Archibald Sinclair. With many traditional domestic Liberal policies now regarded as irrelevant, he focused the party on opposition to both the rise of Fascism in Europe and the appeasement foreign policy of the British government, arguing that intervention was needed, in contrast to the Labour calls for pacifism. Despite the party's weaknesses, Sinclair gained a high profile as he sought to recall the Midlothian Campaign and once more revitalise the Liberals as the party of a strong foreign policy.

In 1940 they joined Churchill's wartime coalition government, with Sinclair serving as Secretary of State for Air, the last British Liberal to hold Cabinet rank office for seventy years. However, it was a sign of the party's lack of importance that they were not included in the War Cabinet. At the 1945 general election, Sinclair and many of his colleagues lost their seats to both Conservatives and Labour, and the party returned just 12 MPs to Westminster.[16] But this was just the beginning of the decline. In 1950, the general election saw the Liberals return just nine MPs.[17] Another general election was called in 1951, and the Liberals were left with just six MPs in parliament; all but one of them were aided by the fact that the Conservatives refrained from fielding candidates in those constituencies.[18]

In 1957 this total fell to five when one of the Liberal MPs died and the subsequent by-election was lost to the Labour Party, which selected the former Liberal Deputy Leader Lady Megan Lloyd George as its own candidate. The Liberal Party seemed close to extinction. During this low period, it was often joked that Liberal MPs could hold meetings in the back of one taxi.

Liberal revival

Through the 1950s and into the 1960s the Liberals survived only because a handful of constituencies in rural Scotland and Wales clung to their Liberal traditions, whilst in two English towns, Bolton and Huddersfield, local Liberals and Conservatives agreed to each contest only one of the town's two seats. Jo Grimond, for example, who became Liberal leader in 1956, was MP for the remote Orkney and Shetland islands. Under his leadership a Liberal revival began, marked by the Orpington by-election of March 1962 which was won by Eric Lubbock. There, the Liberals won a seat in the London suburbs for the first time since 1935.

The Liberals became the first of the major British political parties to advocate British membership of the European Economic Community. Grimond also sought an intellectual revival of the party, seeking to position it as a non-socialist radical alternative to the Conservative government of the day. In particular he canvassed the support of the young post-war university students and recent graduates, appealing to younger voters in a way that many of his recent predecessors had not, and asserting a new strand of Liberalism for the post-war world.

The new middle-class suburban generation began to find the Liberals' policies attractive again. Under Grimond (who retired in 1967) and his successor, Jeremy Thorpe, the Liberals regained the status of a serious third force in British politics, polling up to 20% of the vote but unable to break the duopoly of Labour and Conservative and win more than fourteen seats in the Commons. An additional problem was competition in the Liberal heartlands in Scotland and Wales from the Scottish National Party and Plaid Cymru who both grew as electoral forces from the 1960s onwards. Although Emlyn Hooson held on to the seat of Montgomeryshire, upon Clement Davies death in 1962, the party lost its other Welsh five Welsh seats between 1950 and 1966. In September 1966 the Welsh Liberal Party formed their own state party, moving the Liberal Party into a fully federal structure.[19]

In local elections Liverpool remained a Liberal stronghold, with the party taking the plurality of seats on the elections to the new Liverpool Metropolitan Borough Council in 1973. In the February 1974 general election the Conservative government of Edward Heath won a plurality of votes cast, but the Labour Party gained a plurality of seats due to the Ulster Unionist MPs refusing to support the Conservatives after the Northern Ireland Sunningdale Agreement. The Liberals now held the balance of power in the Commons. Conservatives offered Thorpe the Home Office if he would join a coalition government with Heath. Thorpe was personally in favour of it, but the party insisted on a clear government commitment to introducing proportional representation and a change of Prime Minister. The former was unacceptable to Heath's Cabinet and the latter to Heath personally, so the talks collapsed. Instead a minority Labour government was formed under Harold Wilson but with no formal support from Thorpe. In the October 1974 general election the Liberals slipped back slightly and the Labour government won a wafer-thin majority.

Thorpe was subsequently forced to resign after allegations about his private life. The party's new leader, David Steel, negotiated the Lib-Lab pact with Wilson's successor as Prime Minister, James Callaghan. According to this pact, the Liberals would support the government in crucial votes in exchange for some influence over policy. The agreement lasted from 1977 to 1978, but proved mostly fruitless, for two reasons: the Liberals' key demand of proportional representation was rejected by most Labour MPs, whilst the contacts between Liberal spokespersons and Labour ministers often proved detrimental, such as between finance spokesperson John Pardoe and Chancellor of the Exchequer Denis Healey, who were mutually antagonistic.

Alliance with Social Democrats

When the Labour government fell in 1979, the Conservatives under Margaret Thatcher won a victory which served to push the Liberals back into the margins.

In 1981, defectors from the moderate wing of the Labour Party, led by former Cabinet ministers Roy Jenkins, David Owen and Shirley Williams, founded the Social Democratic Party. The new party and the Liberals quickly formed an alliance, which for a while polled as high as 50% in the opinion polls and appeared capable of winning the next general election. However, they were later overtaken in the polls by the Conservatives and at the 1983 general election the Conservatives triumphed by a landslide, with Labour once again forming the opposition, while the SDP-Liberal Alliance came close to Labour in terms of votes (a share of more than 25%) although it only had 23 MPs compared to Labour's 209.

The 1987 election saw the Alliance's share of the votes drop slightly and it now had 22 MPs, and in the election's aftermath Liberal leader David Steel proposed a merged of the two parties. Most SDP members voted in favour of the merger, but SDP leader David Owen objected and continued to lead a "rump" SDP, with the merger of the two parties being completed in March 1988 to form the **Social and Liberal Democrats**, becoming the **Liberal Democrats** in October 1989.[20]

Merger with Social Democrats

In 1988 the Liberals and Social Democrats merged to create what came to be called the Liberal Democrats. Over two-thirds of the members, and all the serving MPs, of the Liberal Party joined this party, led first jointly by Steel and the SDP leader Robert Maclennan, and later by Paddy Ashdown (1988–99), Charles Kennedy (1999–2006), Sir Menzies Campbell (2006–07) and Nick Clegg (incumbent).

Though the merger process was traumatic and the new party suffered a few years of extremely poor poll results, it gradually found much greater electoral success than the Liberal Party had done in the post-war era. In the 2005 general election, the Liberal Democrats elected 62 MPs to the House of Commons, a far cry from the days when the Liberals had just 5 MPs and Liberalism as a political force had seemed moribund.

As was the case with the Liberal Party for most of the 20th century, the Liberal Democrats face constant questioning about which of the other two parties they are closer to, in particular about which they would support in the event of a hung parliament. The party is keen to maintain its independent identity however, and argues that the need for a modern Liberal force in British politics has never been greater.

In the 2010 General Election, the Conservative Party won more seats than any other, but not enough to form a majority government. After several days of negotiation, the Liberal Democrats agreed to join the Conservatives as part of a coalition government.

The post-1988 Liberal Party

A group of Liberal opponents of the merger with the Social Democrats, including Michael Meadowcroft (formerly Liberal MP for Leeds West) and Paul Wiggin (who served on Peterborough City Council as a Liberal), continued under the old name of "the Liberal Party". This was legally a new organisation (the headquarters, records, assets and debts of the old party were inherited by the Liberal Democrats), but its constitution asserts it to be the same Liberal party. The party retains influence in some local councils. Meadowcroft himself eventually joined the Liberal Democrats in 2007.

National Liberal Party

In 1931 a number of Liberal MPs split from the original Liberal Party in protest at the leadership's adherence to free trade. The party was reformed in 2006 by the political think tank the Third Way under the name 'National Liberal Party'.[21]

Liberal leaders 1859–1988

Liberal Leaders in the House of Lords, 1859–1988

- Granville George Leveson-Gower, 2nd Earl Granville 1859–1865
- John Russell, 1st Earl Russell 1865–1868
- Granville George Leveson-Gower, 2nd Earl Granville 1868–1891
- John Wodehouse, 1st Earl of Kimberley 1891–1894
- Archibald Philip Primrose, 5th Earl of Rosebery 1894–1896
- John Wodehouse, 1st Earl of Kimberley 1896–1902
- John Spencer, 5th Earl Spencer 1902–1905
- George Robinson, 1st Marquess of Ripon 1905–1908
- Robert Crewe-Milnes, 1st Marquess of Crewe 1908–1923
- Edward Grey, 1st Viscount Grey of Fallodon 1923–1924
- William Lygon, 7th Earl Beauchamp 1924–1931
- Rufus Isaacs, 1st Marquess of Reading 1931–1936
- Robert Crewe-Milnes, 1st Marquess of Crewe 1936–1944
- Herbert Samuel, 1st Viscount Samuel 1944–1955
- Philip Rea, 2nd Baron Rea 1955–1967
- Frank Byers, Baron Byers 1967–1984
- Beatrice Seear, Baroness Seear 1984–1989

Liberal Leaders in the House of Commons, 1859–1916

- Henry John Temple, 3rd Viscount Palmerston 1859–1865
- William Ewart Gladstone 1865–1875
- Spencer Compton Cavendish, Marquess of Hartington 1875–1880
- William Ewart Gladstone 1880–1894
- Sir William Vernon Harcourt 1894–1898
- Sir Henry Campbell-Bannerman 1899–1908
- Herbert Henry Asquith, 1st Earl of Oxford and Asquith (1925) 1908–1916

Leaders of the Liberal Party, 1916–1988

- Herbert Henry Asquith, 1st Earl of Oxford and Asquith (1925) 1916–1926
  - *Donald Maclean, Acting Leader* 1919–1920
- David Lloyd George 1926–1931
- Sir Herbert Samuel 1931–1935
- Sir Archibald Sinclair 1935–1945
- Clement Davies 1945–1956
- Jo Grimond 1956–1967
- Jeremy Thorpe 1967–1976
- Jo Grimond 1976 (Acting Leader)
- David Steel 1976–1988

Deputy Leaders of the Liberal Party in the House of Commons

- Herbert Samuel 1929–1931
- Archibald Sinclair 1931–1935
- *Post vacant* 1935–1945
- Percy Harris 1940–1945
- *Post vacant* 1945–1949
- Megan Lloyd George 1949–1951
- *Post vacant* 1951–1962
- Donald Wade 1962–1964
- *Post vacant* 1964–1979
- John Pardoe 1976–1979
- *Post vacant* 1979–1985
- Alan Beith 1985–1988

Deputy Leaders of the Liberal Party in the House of Lords

- Eric Drummond, 16th Earl of Perth 1946–1951
- Walter Layton, 1st Baron Layton 1952–1955
- *Post vacant* 1955–1965
- Gladwyn Jebb, 1st Baron Gladwyn 1965–1988

See also

- Category:Liberal MPs (UK)
- List of Liberal Party (UK) MPs
- Contributions to liberal theory
- Liberalism worldwide
- List of liberal parties
- Liberal democracy
- Liberalism in the United Kingdom
- List of United Kingdom Liberal Party Leaders
- List of Liberal Chief Whips
- President of the Liberal Party
- Politics of the United Kingdom
- Liberalism
- List of liberal theorists
- List of UK Liberal Party general election manifestos

Notes

[1] *The Times* (31 December 1872), p. 5.

[2] W. H. Greenleaf, *The British Political Tradition. Volume II: The Ideological Heritage* (London: Methuen, 1983), p. 143.

[3] "Education Scotland – Higher Bitesize Revision – History – Liberal – Impact: Revision 1" (http://www.bbc.co.uk/scotland/education/bitesize/higher/history/liberal/impact1_rev.shtml). BBC. . Retrieved 2010-04-25.

[4] "The Liberal reforms 1906–1914" (http://www.bbc.co.uk/schools/gcsebitesize/history/mwh/britain/liberalreformsrev1.shtml). *GCSE Bitesize*. BBC. . Retrieved 2009-02-28. "One government that is often seen as an example of 'reforming' by introducing positive changes that really improve peoples' lives is the Liberal government in Britain of 1906–1914. Many historians label this period the beginning of the welfare state [...]"

[5] Liberal Industrial Inquiry, *Britain's Industrial Future* (London, 1928), p. 453.

[6] "1964 Liberal Party Election Manifesto" (http://www.libdemmanifesto.com/1964/1964-liberal-manifesto.shtml). Liberal / SDP / Libdem Manifestos. . Retrieved 2010-04-25.

[7] J. P. Parry, *Democracy and Religion: Gladstone and the Liberal Party, 1867–1875* (1989) p 174
[8] H. C. G. Matthew, "Gladstone, William Ewart (1809–1898)", *Oxford Dictionary of National Biography* (2004)
[9] J. L. Hammond and M. R. D. Foot, *Gladstone and Liberalism* (1952)
[10] Graham D. Goodlad, *British foreign and imperial policy, 1865–1919* (2000) p. 21
[11] Pearce, Robert and Stearn, Roger (2000). ***Access to History**, Government* and *Reform: Britain 1815–1918 (Second Edition), page 56–7.* Hodder & Stoughton.
[12] Graham D. Goodlad, "The Liberal Party and Gladstone's Land Purchase Bill of 1886. Historical Journal 1989 32(3): 627–641 in JSTOR (http://www.jstor.org/stable/2639536)
[13] Pearce, Robert and Stearn, Roger (2000). ***Access to History**, Government* and *Reform: Britain 1815–1918 (Second Edition), page 74.* Hodder & Stoughton.
[14] Table 2.01 "Summary Results of General Elections 1832–2005 (UK)", *British electoral facts, 1832–2006*, by Colin Rallings and Michael Thrasher, 7th edition, 2007, ISBN 978-0-7546-2712-8, p. 59.
[15] "Election 2010 Results" (http://news.bbc.co.uk/1/shared/election2010/results/). BBC News. .
[16] "1945: Labour landslide buries Churchill" (http://news.bbc.co.uk/hi/english/static/vote2001/in_depth/election_battles/1945_over.stm). *The election battles 1945–1997* (BBC News). .
[17] "1950: Labour limps home" (http://news.bbc.co.uk/hi/english/static/vote2001/in_depth/election_battles/1950_over.stm). *The election battles 1945–1997* (BBC News). .
[18] "1951: Churchill's return" (http://news.bbc.co.uk/hi/english/static/vote2001/in_depth/election_battles/1951_over.stm). *The election battles 1945–1997* (BBC News). .
[19] Russell Deacon (2011) History of the Welsh Liberal party, Welsh Academic Press
[20] "SDP: Breaking the mould" (http://news.bbc.co.uk/1/hi/uk_politics/1136223.stm). *BBC News*. 25 January 2001. .
[21] "National Liberal Party Official Site" (http://nationalliberal.org/). .

References

- Chris Cook, *A Short History of the Liberal Party, 1900–2001* (6th edition). Basingstoke: Palgrave, 2002. ISBN 0-333-91838-X.
- Jonathan Parry, *The Rise and Fall of Liberal Government in Victorian Britain*. Yale, 1993.ISBN 0-300-06718-6.
- David Dutton, *A History of the Liberal Party in the 20th Century*, Palgrave Macmillan, 2004. ISBN 0-333-74656-2.

External links

- Liberal Democrat History Group (http://www.liberalhistory.org.uk/)
- Catalogue (http://archives.lse.ac.uk/TreeBrowse.aspx?src=CalmView.Catalog&field=RefNo&key=LIBERAL PARTY) of the Liberal Party papers (mostly dating from after 1945) at LSE Archives (http://www.lse.ac.uk/library/archive/Default.htm)

Social_Democratic_Party_(UK)

Social Democratic Party	
Founded	1981
Dissolved	1988
Split from	Labour Party
Merged into	Social and Liberal Democrats
Succeeded by	Liberal Democrats, Social Democratic Party
Ideology	Centrism, Social democracy
Colours	Blue and red

The **Social Democratic Party** (SDP) was a centrist[1] [2] [3] [4] political party in the United Kingdom that was created on 26 March 1981 and existed until 1988. It was founded by four senior Labour Party 'moderates', dubbed the 'Gang of Four': Roy Jenkins, David Owen, Bill Rodgers and Shirley Williams. At the time of the SDP's creation, Owen and Rodgers were sitting Labour Members of Parliament (MPs); Jenkins had left Parliament in 1977 to serve as President of the European Commission, while Williams had lost her seat in the 1979 general election. The four left the Labour Party as a result of policy changes enacted at the January 1981 Wembley conference which committed the party to unilateral nuclear disarmament and withdrawal from the European Economic Community. They also believed that Labour had become too left-wing, and had been infiltrated at constituency level by Trotskyist factions whose views and behaviour they considered to be at odds with the Parliamentary Labour Party and Labour voters.

For the 1983 and 1987 General Elections, the SDP joined the Liberal Party in the SDP-Liberal Alliance. After a ballot of members and the passing of a motion at the 1987 Portsmouth conference, the party merged with the Liberal Party in 1988 to form the Liberal Democrats, although a minority left to form a continuing SDP led by David Owen.

History

Origins

The origin of the party can be traced back to the ideological divisions in the Labour Party in the 1950s (with its forerunner being the Campaign for Democratic Socialism established to support the Gaitskellites), but publicly lies in the 1979 Dimbleby Lecture given by Roy Jenkins as he neared the end of his presidency of the European Commission. Jenkins argued the necessity for a realignment in British politics, and discussed whether this could be brought about from within the existing Liberal Party, or from a new group driven by European principles of social democracy.

There were long-running claims of corruption and administrative decay within Labour at local level (the North-East of England was to become a cause célèbre), and concerns that experienced and able Labour MPs could be deselected (i.e., lose the Labour Party nomination) by those wanting to put into a safe seat their friends, family or members of their own Labour faction. In some areas, the Militant tendency were held to be systematically targeting weak local party branches in safe seat areas in order to have their own candidates selected, and thus become MPs.

Eddie Milne at Blyth (Northumberland) and Dick Taverne in Lincoln were both victims of such intrigues during the 1970s, but in both cases there was enough of a local outcry by party members – and the electorate – for them to fight and win their seats as independent candidates against the official Labour candidates.

March 1973 Lincoln by-election

In Taverne's case, he had been fighting efforts by the Lincoln Constituency Labour Party to deselect him largely over his support for British membership of the European Communities. In October 1972 he resigned his seat to force a by-election in which he fought as a Democratic Labour candidate against the official party candidate. Taverne won by an unexpectedly large margin.[5] He founded the short lived Campaign for Social Democracy (CFSD) thereafter, and wrote a book about events surrounding the by-election called *The Future of the Left – Lincoln and After* (1972). But the CFSD failed to gain nationwide support, and Taverne lost the seat at the October 1974 General Election. Some independent Social Democrats contested the October 1974 and 1979 General Elections, but none were elected.

Taverne's Lincoln by-election campaign was also helped to a lesser degree by problems with the Conservative candidate, Monday Club chairman Jonathan Guinness. His suggestion during the by-election that murderers should have razor blades left in their cells so they could decently commit suicide resulted in him being nicknamed "Old Razor Blades" during the campaign. This, combined with considerable Conservative grassroots disquiet over the Club's links to the National Front, persuaded some Conservative voters to switch to Taverne in protest as much as tactically to ensure Labour suffered an embarrassing loss. (Guinness had been elected as Chairman specifically to eradicate such links.)

The Manifesto Group and the split from Labour

Many original members of the future Social Democratic Party had been members of the Manifesto Group within the Labour Party. This group opposed what they saw as a leftward shift in Labour policy, the increasing prominence within the party of Tony Benn, and the involvement of trade unions in choosing the leader of the Labour Party. They argued that a new type of political force was needed to challenge the Conservative Party. Further, they opposed the creation of an electoral college to elect the leader of the Party, who had previously been elected by members of the Parliamentary Labour Party – in particular, the arrangement of block voting by constituency parties and trade unions, with the total votes of a constituency party or trade union being given to a candidate based on a first-past-the-post within that CLP or union, or changed at the discretion of delegates (similar to primary elections in the United States). They were also vehemently opposed to unilateral nuclear disarmament, an increasingly popular policy amongst members of the party.

The final straw for many in the Manifesto Group was the behaviour of former Chancellor of the Exchequer Denis Healey at a meeting with them during the Labour leadership campaign to replace James Callaghan. He bluntly told those assembled to vote for him and answered their questions uninformatively. At the end, one asked him why they should vote for him, and Healey answered "You have nowhere else to go" (to stop the left-winger Michael Foot from winning). Healey's arrogance convinced many that their days as members of the Labour Party were now over. Ivor Crewe and Anthony King found five defectors who claimed to have voted for Foot in order to saddle Labour with an unelectable leader and make life easier in their new party. One defector, Mike Thomas, said he was tempted to send a telegraph to Healey reading "Have found somewhere else to go".

Newspapers of the period reported that the announcement of the new party came as a complete shock to MPs from all sides of the Commons, including members of the Manifesto Group, as the 'Gang of Four' had kept their preparations a closely guarded secret. One notable Manifesto Group exception was its secretary, future Defence Secretary George Robertson, who was the only officer to remain. The story got around that he had refused to join the new party because he feared he would not be able to keep his Hamilton seat at a general election; local Scottish National Party supporters nicknamed him "Chicken George".

Creation of the SDP

The founding members or 'Gang of Four' were Roy Jenkins, David Owen, Bill Rodgers and Shirley Williams, all leading figures within the Labour Party, and all 'moderates'. They announced the new party at a press conference, after outlining their policies in what became known as the Limehouse Declaration.

Democratic, Democratic Labour, and Radical were all mentioned as possible names for the new party, as well as New Labour (which future Labour leader Tony Blair would use to promote the Labour Party more than a decade later)[6] but eventually Social Democratic was settled on because the 'Gang of Four' consciously wanted to mould the philosophy and ideology of the new party on the Social Democracy practised on mainland Europe.

The opening statement of principles contained in the preamble of the party's constitution stated that "The SDP exists to create and defend an open, classless and more equal society which rejects prejudices based upon sex, race, colour or religion". The constitution set out the establishment of a "Council for Social Democracy" (CSD) which was, in effect, the party's standing conference. Each area party was entitled to elect delegates to the CSD. A number of internal groups flourished within the new party, the most notable of which was the Tawney Society (mirroring the function of the Fabian Society within the Labour Party).

Twenty-eight Labour MPs eventually joined the new party, along with one member of the Conservative Party, Christopher Brocklebank-Fowler. Williams and Jenkins were not at the time MPs, but were elected to the Commons in by-elections at Crosby and Glasgow Hillhead respectively. Much of the party's initial membership came from the Social Democratic Alliance. The party also received a boost with the recruitment of former student leaders from outside the Labour Party. These included former Communist Party of Great Britain member Sue Slipman as well as Conservative party members including Adair Turner, Anna Soubry and Tom Hayhoe.[7]

Although the SDP was seen as being largely a breakaway from the right wing of the Labour Party, an internal party survey found that 60% of its members had not belonged to a political party before, with 25% being drawn from Labour, 10% from the Conservatives and 5% from the Liberals.

The party enjoyed a considerable honeymoon period with the press, who made much mileage out of their quirk for proffering claret at their functions. Claret is an "agreeable" wine, and a metaphor for the party's harmonious internal relations compared to those of the strife-torn Labour Party of the period.

The policies of the SDP emphasised a middle position between perceived extremes of Thatcherism and the Labour Party. Its constitution argued for "the fostering of a strong public sector and a strong private sector without frequent frontier changes". The SDP favoured some Thatcherite reforms during the 1980s, such as legislation aimed at reforming the trade unions (although the parliamentary SDP actually split three ways on Norman Tebbit's 1982 Industrial Relations bill, most voting for, some against, and others abstaining), but took a more welfarist position than the Conservative Party, being more sceptical of Conservative welfare reforms (particularly regarding the Health Service).[8]

At the party's first electoral contest, Jenkins narrowly failed to win a by-election at Warrington in July 1981, describing it as his "first defeat, but by far (his) greatest victory". In the Glasgow Hillhead by-election in March 1982, another candidate named Roy Jenkins was nominated by Labour Party activists to contest the seat in order to confuse voters and split his potential vote. SDP polling agents were given special dispensation by the Returning Officer to have placards outside of polling stations to state which one on the ballot papers was the 'real Roy'. Ultimately, the SDP's Jenkins was elected.

A leadership election was held later in the year, Jenkins beating Owen in the ballot to become the first party leader. Later in the year, Shirley Williams defeated Bill Rogers in the ballot to become SDP president.

The Alliance

The SDP formed the SDP-Liberal Alliance with the Liberal Party late in 1981, under the joint leadership of Roy Jenkins (SDP) and Liberal leader David Steel. The Liberal Party, and in particular its leader, David Steel, had applauded the formation of the SDP from the sidelines from the very start. Senior Liberal MP for Rochdale Cyril Smith caused some embarrassment, however, by publicly stating that the SDP "should be strangled at birth".[9] During an era of public disillusionment with the two main parties – Labour and the Conservatives – and widescale unemployment, the Alliance achieved considerable success in parliamentary by-elections. At one point in late 1981, the party had an opinion poll rating of over 50%.[10]

Also in 1981, David Steel was able to address the Liberal Party conference with the phrase "Go back to your constituencies, and prepare for government!"[11]

In early 1982, after public disagreements over who could fight which seats in the forthcoming election, the poll rating dipped, but the party remained ahead of both Labour and the Conservatives. However, following the outbreak of the Falklands War on 2 April 1982, the Conservative government of Margaret Thatcher soared from third to first place in the public opinion polls. The standing of the Alliance and Labour declined. By this stage, however, the Alliance already 30 MPs in parliament - virtually all of them defectors from Labour, joined by just the one Tory MP.

Labour lost Bermondsey, one of their ten safest seats, in a by-election in February 1983 to Liberal candidate Simon Hughes: the sitting Labour MP Robert Mellish resigned to work for the London Docklands Development Corporation but, being opposed to the selection by his left-wing Constituency Labour Party of Peter Tatchell, supported the former leader of Southwark council John O'Grady as "Real Bermondsey Labour" giving an impression of Labour division and infighting.

In the 1983 general election, the SDP/Liberal Alliance won more than 25% of the national vote, close behind Labour's 28%, but well behind the 44% secured by the Conservatives. However, because of the British "first-past-the-post" electoral system, only 23 Alliance MPs were elected, six of whom were members of the SDP. The party's leader, Roy Jenkins, managed to hold his seat at Glasgow Hillhead, but SDP President Shirley Williams was defeated at Crosby (which she had won in a by-election in November 1981) as a result of unfavourable boundary changes. Labour leader Michael Foot, who resigned within days of the election, was critical of the SDP/Liberal Alliance for siphoning support away from Labour, allowing the Conservatives to win more seats and secure a triple-digit majority, while Labour was left with 209 seats in parliament.[12]

The MP for Plymouth Devonport, Dr David Owen (who had been a Labour government minister under Harold Wilson and James Callaghan between 1974 and 1979), took over as SDP leader after the 1983 General Election. He was more sceptical about close relations with the Liberals than his predecessor Roy Jenkins, and favoured retaining the party's distinct identity. Owen's influence ensured that proposals for a merger between the two parties were shelved after a lengthy debate at the SDP's 1983 Conference.

During the 1983-87 parliament some SDP members started to become unsettled at what appeared to be the increasingly right wing course taken by SDP leader David Owen. This resulted in some members launching the Limehouse Group in an attempt to keep the party on the left of centre course that was first propounded in the Limehouse Declaration.

Two more SDP MPs were elected in by-elections during the 1983-87 parliament, but in the 1987 general election, the Alliance's share of the vote fell to 23%, and the SDP's parliamentary party was reduced from eight members to five. Roy Jenkins was amongst those who lost their seats. Mike Hancock had won a by-election at Portsmouth South in 1984 from the Conservatives which was lost in 1987, but Rosie Barnes, who had won the bitterly contested Greenwich by-election in February 1987 from Labour managed to hold on in the June General Election.

From the outset, the formation of the Alliance had raised questions as to whether it would lead to a merged party, or the two parties were destined to compete with each other. This in turn led to grassroots tensions in some areas between Liberal and SDP branches that impaired their ability to mount joint campaigns successfully. Such cross-party feuding was part of the reason for Jenkins losing his Hillhead seat to Labour candidate George Galloway in 1987.

Liberal pride was damaged by the sustained lampooning of the Alliance by ITV's *Spitting Image* puppet comedy programme portraying Steel as the craven lickspittle of Owen. One *Spitting Image* sketch had a Machiavellian Owen proposing to a simpering Steel that the parties merged under a new name: "and for our side we'll take 'Social Democratic', and from your side, we'll take 'Party'", to which a hesitant Steel agreed.

Merger, disestablishment and splits

After the disappointment of 1987, Steel proposed a formal merger of the two parties. Jenkins and Steel had believed this to be eventually inevitable after the party failed to break through at the 1983 election. The proposal, also supported by Williams and Rodgers, was fiercely opposed by Owen, who argued that such a merger would not be accepted by the electorate, and would not reverse their declining share of the vote. Jenkins denied that a merger had been his original intent.

But the majority of the SDP's membership (along with those of the Liberals) voted in favour of the union. Owen resigned as leader and was replaced by Robert Maclennan. Steel and Maclennan headed the new "Social and Liberal Democrat Party" from 3 March 1988. An interim working name for the party, the "Democrats", was adopted by conference on 26 September 1988. This proved to be unpopular, and the party was renamed the Liberal Democrats in October 1989, as had been originally proposed at the September 1988 conference by the party's Tiverton branch.

Most SDP members, including SDP MP and future Liberal Democrat leader Charles Kennedy, joined Maclennan in the merged party, but Owen created a continuing SDP, along with two other MPs, John Cartwright and Rosie Barnes. There was also a continuing Liberal Party, led by Michael Meadowcroft and David Morrish, mainly based on Liverpool and West Country Liberals who feared a dilution by the former SDP members of the Liberal tradition within the merged party.

Some members simply dropped out of politics altogether out of disillusionment when the time came to renew their membership subscriptions. After a series of highly publicised expulsions of Militant tendency members, the Labour Party led by Neil Kinnock benefited in the opinion polls from the feuding within the former Alliance. The Liberal Democrats also lost to the continuing SDP its one major backer, Lord Sainsbury.

The subsequent election of a new Liberal leader, Paddy Ashdown, revived the new party's fortunes in time, and turned it into the most successful "third party" electorally in British politics since the days of Lloyd George. Under David Owen, the SDP continued from 1988 to 1990, and subsequently, but without Owen's involvement.

In 1996, the SDP vowed to continue as a party, with their new leader, John Bates, who continued until 1999, when a new interim leader was elected, until 2006, when John Bates returned to the party leader. In 2008, Peter Johnson became leader, and changed its position on Europe, adopting as its chief policy of withdrawal from the European Union. The party continues as a non-parliamentary party. Its strongest base is Bridlington.[13]

Aftermath

Most members of the SDP who joined the Liberal Democrats have remained in that party. There have been a few exceptions to this, with Roger Liddle and Polly Toynbee among those who former Labour Party members of the SDP who returned to Labour ranks subsequently. Roger Liddle went on to become Special Adviser on European matters to Tony Blair and Chairman of the international think tank Policy Network in which capacity he has had a major influence on modernising the political philosophy of the Labour Party. Some Owenites joined the Conservative Party, with one, Danny Finkelstein, becoming a close aide of both John Major and William Hague.

But most important of all, the Social Democratic Party strengthened the political credibility of the Liberals. The national status of Roy Jenkins (former Chancellor and Home Secretary) and David Owen (former Foreign Secretary who had been widely tipped as a future Labour Prime Minister) helped the Liberals become something more than a source of shock by-election results and a party for those living in rural areas such as the Scottish Highlands and the West Country. The SDP also helped the Liberals attract attention from the media for their policies after a long period when the only media interest in the party resulted from the trial of former Liberal leader Jeremy Thorpe.

See also

- Politics of the United Kingdom
- Liberal Democrat History Group

Leaders of the SDP

#	Name (Birth–Death)	Portrait	Constituency	Entered office	Left office
1	Roy Jenkins (1920–2003)		Glasgow Hillhead from 1982	7 July 1981	13 June 1983
2	David Owen (1938–)		Plymouth Devonport	13 June 1983	6 August 1987
3	Robert Maclennan (1936–)		Caithness and Sutherland	29 August 1987	3 March 1988

Further reading

- Ivor Crewe & Anthony King, *SDP: The Birth, Life and Death of the Social Democratic Party*, OUP 1995
- Hugh Stephenson, *Claret and Chips: The Rise of the SDP*, Michael Joseph 1982
- John Stevenson, *Third Party Politics since 1945: Liberals, Alliance and Liberal Democrats*, ICBH 1993
- Patricia Lee Sykes, *Losing from the Inside: The Cost of Conflict in the British Social Democratic Party*, Transaction Publishers 1990
- Roy Jenkins, *A Life at the Centre*, Macmillan 1991
- David Owen, *Time to Declare*, Penguin 1992
- Bill Rogers, *Fourth Among Equals*, Politico's 2000
- Journal of Liberal History, Issue 39 Summer 2003, *A Short History of Political Virginity* (edition devoted to the SDP) http://www.liberalhistory.org.uk/uploads/39-Summer%25202003.pdf
- Journal of Liberal History, Issue 18 Spring 1998, *Ten Years On, The Legacy of the Alliance and Merger* http://www.liberalhistory.org.uk/uploads/18_Spring_1998.pdf

External links

- Policy Network [14]

References

[1] Jeffrey Kopstein; Mark Lichbach (5 September 2005). *Comparative Politics: Interests, Identities, and Institutions in a Changing Global Order* (http://books.google.com/books?id=NDP8RVXvlUMC&pg=PA66). Cambridge University Press. pp. 66–. ISBN 978-0-521-84316-4. . Retrieved 19 August 2012.

[2] Mark Kesselman; Joel Krieger; Christopher S. Allen; Stephen Hellman (12 February 2008). *European Politics in Transition* (http://books.google.com/books?id=FGK0Ht0OM_IC&pg=PA88). Cengage Learning. pp. 88–. ISBN 978-0-618-87078-3. . Retrieved 19 August 2012.

[3] John R. Cook; Peter Wright (6 January 2006). *British Science Fiction Television: A Hitchhiker's Guide* (http://books.google.com/books?id=Bo_vcscFbZEC&pg=PA211). I.B.Tauris. pp. 211–. ISBN 978-1-84511-048-2. . Retrieved 19 August 2012.

[4] Kerstin Hamann; John Kelly (2011). *Parties, Elections, and Policy Reforms in Western Europe: Voting for Social Pacts* (http://books.google.com/books?id=mId4kwMugA4C&pg=PA75). Taylor & Francis. pp. 75–. ISBN 978-0-415-58195-0. . Retrieved 19 August 2012.
[5] John Ramsden and Richard Jay, "Lincoln: Background to Taverne's Triumph" in "By-elections in British Politics", Macmillan, 1973, pp. 264-315.
[6] "The rise and fall of New Labour" (http://www.bbc.co.uk/news/uk-politics-10518842). *BBC News*. 3 August 2010. .
[7] Times; Guardian; Daily Telegraph; Sun; Daily Mail; Daily Express; Mirror 12 - 14 August 1985
[8] Hames, Tim (13 June 2005). "At last it can be told David Owen was Tony Blairs secret political father" (http://www.timesonline.co.uk/tol/comment/columnists/tim_hames/article532682.ece). *The Times* (London). . Retrieved 7 April 2010.
[9] "Home" (http://www.totalpolitics.com/quotations/quotations.php?auth_id=357). Total Politics. . Retrieved 18 April 2010.
[10] "SDP: Breaking the mould" (http://news.bbc.co.uk/1/hi/uk_politics/1136223.stm). *BBC News*. 25 January 2001. .
[11] Stone-Lee, Ollie (10 September 2003). "Conference season's greatest hits" (http://news.bbc.co.uk/1/hi/uk_politics/3185313.stm). *BBC News*. . Retrieved 7 April 2010.
[12] (http://news.bbc.co.uk/onthisday/hi/dates/stories/june/9/newsid_2500000/2500847.stm)
[13] Philpot, Robert (16 January 2006). "The SDP lives on - in Bridlington" (http://www.newstatesman.com/node/152350). *New Statesman*. . Retrieved 10 September 2012.
[14] http://www.policy-network.net/

Commonwealth

Commonwealth is a traditional English term for a political community founded for the common good. Historically, it has sometimes been synonymous with "republic".

More recently it has been used for fraternal associations of some sovereign nations. Most notably, the Commonwealth of Nations, an association primarily of former members of the British Empire, is often referred to as simply "the Commonwealth".

Etymology

The English noun *commonwealth* in the sense meaning "public welfare; general good or advantage" dates from the 15th century.[1] The original phrase "the common-wealth" or "the common weal" (echoed in the modern synonym "public weal") comes from the old meaning of "wealth," which is "well-being", and is itself a loose translation of the Latin res publica (republic). The term literally meant "common well-being." In the 17th century the definition of "commonwealth" expanded from its original sense of "public welfare" or "commonweal" to mean "a state in which the supreme power is vested in the people; a republic or democratic state."[2]

Historic usage

Iceland

The Icelandic Commonwealth or the Icelandic Free State (Icelandic: Þjóðveldið) was the state existing in Iceland between the establishment of the Althing in 930 and the pledge of fealty to the Norwegian king in 1262. It was initially established by a public consisting largely of recent immigrants from Norway who had fled the unification of that country under King Harald Fairhair.

Polish-Lithuanian Commonwealth

Republic is still an alternative translation of the traditional name of the Polish-Lithuanian Commonwealth. Wincenty Kadłubek (Vincent Kadlubo, 1160–1223) used for the first time the original Latin term *res publica* in the context of Poland in his "Chronicles of the Kings and Princes of Poland." The name was used officially for the confederal country formed by Poland and Lithuania 1569–1795.

It is also often referred as "Nobles' Commonwealth" (1505–1795, i.e. before the union). In contemporary political doctrine of Polish-Lithuanian Commonwealth, "our state is a Republic (Commonwealth) under presidency of the

King." The commonwealth introduced a doctrine of religious tolerance called Warsaw Confederation, had its own parliament *Sejm* (although elections were restricted to the nobility and elected kings, who were bound to certain contracts *Pacta conventa* from the beginning of the reign. The foundation stones of the Commonwealth (also called the Golden Freedoms) used to be

- free election of the king
- *Pacta conventa*, a binding pledge agreed to by the King on his election
- *rokosz*, the right of rebellion against kings who did not rule in accordance with their pledge
- *liberum veto* (a later development), the right for a single representative to veto the entire proceedings of the Sejm
- *confœderatio* (confederation), a military organisation of the citizens for the attainment of common political aims.

"A commonwealth of good counsaile" was the title of the 1607 English translation of the work of Wawrzyniec Grzymała Goślicki "De optimo senatore" that presented to English readers many of the ideas present in the political system of the Polish-Lithuanian Commonwealth.

England

The Commonwealth of England was the official name of the political unit (*de facto* military rule in the name of parliamentary supremacy) that replaced the Kingdom of England (after the English Civil War) from 1649 to 1653 and 1659 to 1660. Under the rule of Oliver Cromwell and his son and successor Richard. From 1653 to 1659, although still legally known as a Commonwealth, the republic, united with the former Kingdom of Scotland, operated under different institutions (at times as a *de facto* monarchy) and is known by historians as the Protectorate. The Commonwealth of England formed the first republic in the English-speaking world. In an English context, it is sometimes referred to as the "Old Commonwealth."

National

Australia

The term also served when six Australian colonies federated to form the Commonwealth of Australia in 1901. The Commonwealth of Australia Constitution Act created a federal system, in which power is divided between the federal, or national, government and the States—the evolved status of the colonies. The Constitution stipulated that Australia was a constitutional monarchy, where the Head of State is the British (or, since 1942, Australian) monarch, who is represented at the federal level by a Governor-General, and at the state level by six Governors, one for each state. The Parliament of Australia was derived from the British and American systems to form a uniquely Australian system. It is largely based on the British Westminster System, adopting many of its practices and precedents, but with a similar structure—House of Representatives, and Senate—to the U.S. Congress. In an Australian context, the term "Commonwealth" (capitalised) thus refers to the federal government and "Commonwealth of Australia" is the official name of the country.

Dominica

The small Caribbean republic of Dominica has used the official style *Commonwealth of Dominica* since 1970.

The Bahamas

The Bahamas, a Commonwealth realm, uses the official style *Commonwealth of The Bahamas*.

United States

U.S. states

Four states in the United States officially designate themselves as "commonwealths". All four were original colonies (Kentucky was originally a part of the land grant of the Colony of Virginia) and share a strong influence of colonial common law in some of their laws and institutions. The four are:

- **Kentucky** is designated a Commonwealth by the Kentucky Constitution and is known constitutionally as the "Commonwealth of Kentucky".[3]
- **Massachusetts** is a Commonwealth,[4] declaring itself as such in its constitution, which states that "The body politic is formed by a voluntary association of individuals: it is a social compact, by which the whole people covenants with each citizen, and each citizen with the whole people, that all shall be governed by certain laws for the common good."[5]
- **Pennsylvania** uses the "Commonwealth of Pennsylvania" constitutionally and in its official title.[6]
- **Virginia** has been known as the "Commonwealth of Virginia" since before joining the United States, and is referred to as a Commonwealth in its constitution.[7]

U.S. insular areas

"Commonwealth" is also used in the U.S. to describe the political relationship between the United States and the overseas unincorporated territories:

- **Commonwealth of the Philippines** — formed in 1934, and became independent in 1946, upon which "Commonwealth" was replaced by "Republic".
- **Commonwealth of Puerto Rico** — became a commonwealth[8] in 1952.
- **Northern Mariana Islands** — became a commonwealth in 1978.

United Kingdom

Commonwealth of Britain Bill

Labour MP Tony Benn introduced a Commonwealth of Britain Bill several times between 1991 and 2001, intended to abolish the monarchy and establish a British republic. It never reached second reading.

International

Commonwealth of Nations

The Commonwealth of Nations—formerly the British Commonwealth—is a voluntary association of 54 independent sovereign states, most of which are former British colonies, or dependencies of these colonies with three exceptions, Mozambique (which was a Portuguese possession), Rwanda (which was a Belgian mandate) and Cameroon[9] (which is a union of a French mandate and a British mandate) plus the United Kingdom itself. The Commonwealth's membership includes both republics and monarchies. The head of the Commonwealth of Nations is Queen Elizabeth II. She also reigns as monarch directly in a number of states, known as Commonwealth realms.

Commonwealth of Independent States

The Commonwealth of Independent States (CIS) is a loose alliance or confederation consisting of 10 of the 15 former Soviet Republics, the exceptions being Turkmenistan (a CIS associate member), Lithuania, Latvia, Estonia and Georgia. Georgia left the CIS in August 2008 after a clash with Russia over South Ossetia. Its creation signaled the dissolution of the Soviet Union, its purpose being to "allow a civilized divorce" between the Soviet Republics. The CIS has developed as a forum by which the member-states can co-operate in economics, defense and foreign policy.

See also

- Co-operative Commonwealth (society)
- League
- Confederation
- Federation

References

[1] "Commonwealth" (http://dictionary.oed.com/cgi/entry/50045157?query_type=word&queryword=commonwealth&first=1&max_to_show=10&sort_type=alpha&result_place=1&search_id=Hr0v-iyV0EE-153&hilite=50045157), *Oxford English Dictionary* (dictionary.oed.com), 1989, 2nd ed., , retrieved 13 March 2010

[2] "Better things were done, and better managed ... under a Commonwealth than under a King." Pepys, *Diary* (1667) "Commonwealth" (http://dictionary.oed.com/cgi/entry/50045157?query_type=word&queryword=commonwealth&first=1&max_to_show=10&sort_type=alpha&result_place=1&search_id=Hr0v-iyV0EE-153&hilite=50045157), *Oxford English Dictionary* (dictionary.oed.com), 1989, 2nd ed., , retrieved 13 March 2010

[3] http://kentucky.gov

[4] http://www.mass.gov

[5] http://www.malegislature.gov/Laws/Constitution#cp00s00.htm

[6] http://www.pa.gov

[7] http://www.virginia.gov

[8] http://welcome.topuertorico.org/government.shtml: copyright 2009

[9] http://www.thecommonwealth.org/YearbookInternal/145151/history/

External links

- Commonwealth of Nations
  - The Commonwealth (http://www.direct.gov.uk/Gtgl1/GuideToGovernment/InternationalBodies/InternationalBodiesArticles/fs/en?CONTENT_ID=4003092&chk=JXlogH) — UK government site
  - Commonwealth of Nations Secretariat (http://www.thecommonwealth.org/)*
  - Commonwealth Foundation (http://www.commonwealthfoundation.com/)
  - Royal Commonwealth Society (http://www.rcsint.org/)
- Commonwealth of Independent States
  - CIS Executive Committee (http://www.cis.minsk.by/)
  - CIS Statistical Committee (http://www.cisstat.com/eng/)
- Countries
  - Commonwealth of Australia (http://www.gov.au/)
- United States of America
  - Commonwealth of Kentucky (http://www.ky.gov)
  - Commonwealth of Massachusetts (http://www.mass.gov)
  - Commonwealth of Pennsylvania (http://www.state.pa.us/)
  - Commonwealth of Virginia (http://www.virginia.gov/cmsportal/)

  - Commonwealth of Puerto Rico (http://www.gobierno.pr/)
  - Commonwealth of the Northern Mariana Islands (http://www.gov.mp/)
- Polish-Lithuanian Commonwealth
  - Commonwealth of Diverse Cultures: Poland's Heritage (http://www.commonwealth.pl)

Concorde

Concorde	

Role	Supersonic airliner
Manufacturer	BAC (now BAE Systems) Sud-Aviation, Aérospatiale (now EADS)
First flight	2 March 1969
Introduction	21 January 1976
Retired	26 November 2003
Status	Retired from service
Primary users	British Airways Air France See *Operators* below for others
Number built	20 (including 6 non-airline aircraft)[1] [2]
Program cost	£1.3 billion[3]
Unit cost	£23 million in 1977

Aérospatiale-BAC Concorde (pron.: /ˈkɒŋkɔrd/) is a retired turbojet-powered supersonic passenger airliner or supersonic transport (SST). It is the second of only two SSTs to have entered commercial service; the first was the Tupolev Tu-144. Concorde was a product of the manufacturing efforts of Aérospatiale and the British Aircraft Corporation (BAC), produced under a joint Franco-British treaty. First flown in 1969, Concorde entered service in 1976 and continued commercial flights for 27 years.

Among other destinations, Concorde flew regular transatlantic flights from London Heathrow and Paris-Charles de Gaulle Airport to New York JFK and Washington Dulles; it profitably flew these routes in less than half the time of other airliners. With only 20 aircraft built, the development of Concorde was a substantial economic loss; Air France and British Airways also received considerable government subsidies to purchase them. Concorde was retired in 2003 due to a general downturn in the aviation industry after the type's only crash in 2000, the 9/11 terrorist attacks in 2001 and a decision by Airbus, the successor firm of Aerospatiale and BAC, to discontinue maintenance support.[4]

Concorde's name reflects the development agreement between the United Kingdom and France. In the UK, any or all of the type—unusual for an aircraft—are known simply as "Concorde", without an article. The aircraft is regarded by many people as an aviation icon and an engineering marvel.[5]

Development

Concept

In the late 1950s, France, the United Kingdom, United States, and Soviet Union were considering developing supersonic transport. The British Bristol Aeroplane Company and the French Sud Aviation were both working on designs, called the Type 223 and Super-Caravelle, respectively. Both were largely funded by their respective governments.[6] The British design was for a thin-winged delta shape[7] (which owed much to work by Dietrich Küchemann, then at the Royal Aircraft Establishment) for a transatlantic-ranged aircraft for about 100 people, while the French were intending to build a medium-range aircraft.[6]

Concorde's final flight; G-BOAF from Heathrow to Bristol, on 26 November 2003. The extremely high fineness ratio of the fuselage is evident.

The designs were both ready to start prototype construction in the early 1960s, but the cost was so great that the British government made it a requirement that British Aircraft Corporation (which had been formed in 1960 as a consolidation of British aircraft companies, including the Bristol Aeroplane Company) look for international co-operation.[6] Approaches were made to a number of countries, but only France showed real interest. The development project was negotiated as an international treaty between the two countries rather than a commercial agreement between companies and included a clause, originally asked for by the UK, imposing heavy penalties for cancellation. A draft treaty was signed on 29 November 1962. By this time, both companies had been merged into new ones; thus, the Concorde project was between the British Aircraft Corporation and Aérospatiale.[6] At first the new consortium intended to produce one long range and one short range version. However, prospective customers showed no interest in the short-range version and it was dropped.[6] The consortium secured orders (i.e., non-binding options) for over 100 of the long-range version from the major airlines of the day: Pan Am, BOAC, and Air France were the launch customers, with six Concordes each. Other airlines in the order book included Panair do Brasil, Continental Airlines, Japan Airlines, Lufthansa, American Airlines, United Airlines, Air India, Air Canada, Braniff, Singapore Airlines, Iran Air, Olympic Airways, Qantas, CAAC, Middle East Airlines, and TWA.[6] [8] [9]

The design work was supported by a preceding research programme studying the flight characteristics of low ratio delta wings. The supersonic BAC 221 was modified for flight tests of the high speed flight envelope,[10] the Handley Page HP.115 also provided valuable information on low speed performance.[11]

Naming

Reflecting the treaty between the French and British governments which led to Concorde's construction, the name *Concorde* is from the French word *concorde* (IPA: [kɔ̃kɔʁd]), which has an English equivalent, *concord*. Both words mean *agreement*, *harmony* or *union*.

British Airways Concorde in early BA livery at London Heathrow Airport, in the 1980s

The aircraft was initially referred to in the UK as *Concorde*, with the French spelling, but was officially changed to *Concord* by Harold Macmillan in response to a perceived slight by Charles de Gaulle.[12] In 1967, at the French roll-out in Toulouse the British Government Minister for Technology, Tony Benn, announced that he would change the spelling back to *Concorde*.[12] This created a nationalist uproar that died down when Benn stated that the suffixed e represented "Excellence, England, Europe and Entente (Cordiale)." In his memoirs, he recounts a tale of a letter from an irate Scotsman claiming: "[Y]ou talk about 'E' for England, but part of it is made in Scotland." Given Scotland's contribution of providing the nose cone for the aircraft, Benn replied, "[I]t was also 'E' for 'Écosse' (the French name for Scotland) — and I might have added 'e' for extravagance and 'e' for escalation as well!"[13]

Concorde also acquired an unusual nomenclature for an aircraft. In common usage in the United Kingdom, the type is known as *Concorde* without an article, rather than ***the*** *Concorde* or ***a*** *Concorde*.[14] [15]

Testing

Construction of two prototypes began in February 1965: 001, built by Aerospatiale at Toulouse, and 002, by BAC at Filton, Bristol. Concorde 001 made its first test flight from Toulouse on 2 March 1969, piloted by André Turcat,[16] and first went supersonic on 1 October.[17] The first UK-built Concorde flew from Filton to RAF Fairford on 9 April 1969, piloted by Brian Trubshaw.[18] [19] Both prototypes were presented to the public for the first time on 7–8 June 1969 at the Paris Airshow. As the flight programme progressed, 001 embarked on a sales and demonstration tour on 4 September 1971, which was also the first transatlantic crossing of Concorde.[20] [21] Concorde 002 followed suit on 2 June 1972 with a tour of the Middle and Far East.[22] Concorde 002 made the first visit to the United States in 1973, landing at the new Dallas/Fort Worth Regional Airport to mark that airport's opening.[23]

Concorde's first visit to Heathrow Airport on 1 July 1972.

While Concorde had initially held a great deal of customer interest, the project was hit by a large number of order cancellations. The Paris Le Bourget air show crash of the competing Soviet Tupolev Tu-144 had shocked potential buyers, and public concern over the environmental issues presented by a supersonic aircraft – the sonic boom, takeoff-noise and pollution – had produced a shift in public opinion of SSTs. By 1976 four nations remained as prospective buyers: Britain, France, China, and Iran.[24] Only Air France and British Airways (the successor to BOAC) took up their orders, with the two governments taking a cut of any profits made.[25]

The United States cancelled the Boeing 2707, its rival supersonic transport programme, in 1971. Observers have suggested that opposition to Concorde on grounds of noise pollution had been encouraged by the United States Government, as it lacked its own competitor.[26] The US, India, and Malaysia all ruled out Concorde supersonic flights over the noise concern, although some of these restrictions were later relaxed.[27] [28] Professor Douglas Ross characterised restrictions placed upon Concorde operations by President Jimmy Carter's administration as having been an act of protectionism of American aircraft manufacturers.[29]

Concorde had other considerable difficulties that led to its dismal sales performance. Costs had spiralled during development to more than six times the original projections, arriving at a unit cost of £23 million in 1977.[30] World events had also dampened Concorde sales prospects, the 1973 oil crisis had many airlines think twice about aircraft with high rates of fuel-consumption; and new wide-body aircraft, such as the Boeing 747, had recently made subsonic aircraft significantly more efficient and presented a low-risk option for airlines.[31] While carrying a full load, Concorde achieved 15.8 passenger miles per gallon of fuel, while the Boeing 707 reached 33.3 pm/g, the Boeing 747 46.4 pm/g, and the McDonnell Douglas DC-10 53.6 pm/g.[32] An emerging trend in the industry in favour of cheaper airline tickets had also caused airlines such as Qantas to question Concorde's market suitability.[33]

Design

General features

Concorde is an ogival (also "ogee") delta-winged aircraft with four Olympus engines based on those employed in the RAF's Avro Vulcan strategic bomber. Concorde was the first airliner to have a (in this case, analogue) fly-by-wire flight-control system; the avionics of Concorde were unique because it was the first commercial aircraft to employ hybrid circuits.[34] The principal designer for the project was Pierre Satre, with Sir Archibald Russell as his deputy.[35]

Concorde cockpit layout

Concorde pioneered the following technologies:

For high speed and optimisation of flight:

- Double delta (ogee/ogival) shaped wings[7]
- Variable engine air intake system controlled by digital computers[36]
- Supercruise capability[37]
- Thrust-by-wire engines, predecessor of today's FADEC-controlled engines[36]
- Droop-nose section for better landing visibility

For weight-saving and enhanced performance:

- Mach 2.04 (~2179 kilometres per hour / mph) cruising speed[38] for optimum fuel consumption (supersonic drag minimum although turbojet engines are more efficient at higher speed[39])
- Mainly aluminium construction for low weight and conventional manufacture (higher speeds would have ruled out aluminium)[40]
- Full-regime autopilot and autothrottle[41] allowing "hands off" control of the aircraft from climb out to landing
- Fully electrically controlled analogue fly-by-wire flight controls systems[34]
- High-pressure hydraulic system of 28 MPa (4,000 lbf/in²) for lighter hydraulic components[42]
- Complex Air Data Computer (ADC) for the automated monitoring and transmission of aerodynamic measurements (total pressure, static pressure, angle of attack, side-slip).[43]
- Fully electrically controlled analogue brake-by-wire system[44]
- Pitch trim by shifting fuel around the fuselage for centre-of-gravity control[45]
- Parts made using "sculpture milling", reducing the part count while saving weight and adding strength.[46]
- No auxiliary power unit, as Concorde would only visit large airports where ground air start carts are available.[47]

Engines

Close up of pre-production engine nozzles of G-AXDN. The nozzle/thrust reverser design was altered for production

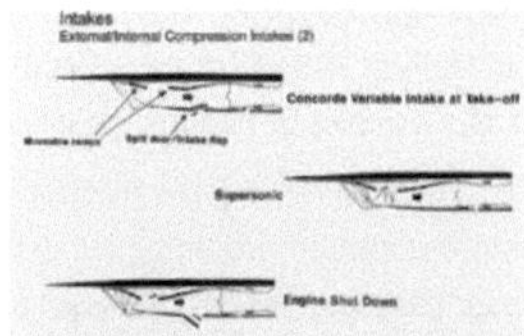

Concorde's intake system schematics

Concorde's intake system

Concorde needed to fly long distances to be economically viable; this required high efficiency. Turbofan engines were rejected due to their larger cross-section producing excessive drag. Turbojets were found to be the best choice of engines.[48] The engine used was the twin spool Rolls-Royce/Snecma Olympus 593, a development of the Bristol engine first used for the Avro Vulcan bomber, and developed into an afterburning supersonic variant for the BAC TSR-2 strike bomber.[49]

The aircraft used reheat (afterburners) at takeoff and to pass through the transonic regime (i.e., "go supersonic") between Mach 0.95 and Mach 1.7. The afterburners were switched off at all other times.[50] Due to jet engines being highly inefficient at low speeds, Concorde burned two tonnes of fuel (almost 2% of the maximum fuel load) taxiing to the runway.[51] To conserve fuel only the two outer engines were run after landing for taxiing.

The intake design for Concorde's engines was especially critical.[52] Conventional jet engines can take in air at only around Mach 0.5; therefore the air has to be slowed from the Mach 2.0 airspeed that enters the engine intake. In particular, Concorde needed to control the shock waves that this reduction in speed generates to avoid damage to the engines. This was done by a pair of intake ramps and an auxiliary spill door, whose position moved in-flight to slow transiting air.[53]

Engine failure causes problems on conventional subsonic aircraft; not only does the aircraft lose thrust on that side but the engine creates drag, causing the aircraft to yaw and bank in the direction of the failed engine. If this had happened to Concorde at supersonic speeds, it theoretically could have caused a catastrophic failure of the airframe. Although computer simulations predicted considerable problems, in practice Concorde could shut down both engines on the same side of the aircraft at Mach 2 without the predicted difficulties.[54] During an engine failure the required air intake is virtually zero so, on Concorde, engine failure was countered by the opening of the auxiliary spill door and the full extension of the ramps, which deflected the air downwards past the engine, gaining lift and minimising drag. Concorde pilots were routinely trained to handle double engine failure.[55]

Heating issues

Air friction on the outer surfaces caused the cabin to heat up during flight. Every surface, such as windows and panels, was warm to the touch by end of the flight.[56] Besides engines, the hottest part of the structure of any supersonic aircraft, due to aerodynamic heating, is the nose. The engineers used Hiduminium R.R. 58, an aluminium alloy, throughout the aircraft due to its familiarity, cost and ease of construction. The highest temperature that aluminium could sustain over the life of the aircraft was 127 °C (°F), which limited the top speed to Mach 2.02.[57] Concorde went through two cycles of heating and cooling during a flight, first cooling down as it gained altitude, then heating up after going supersonic. The reverse happened when descending and slowing down. This had to be factored into the metallurgical and fatigue modelling. A test rig was built that repeatedly heated up a full-size section of the wing, and then cooled it, and periodically samples of metal were taken for testing.[58] [59]

Owing to air friction as the plane traveled at supersonic speed, the fuselage would heat up and expand by as much as 300 mm (almost 1 ft). The most obvious manifestation of this being a gap that opened up on the flight deck between the flight engineer's console and the bulkhead. On some aircraft that conducted a retiring supersonic flight, the flight engineers placed their caps in this expanded gap, wedging the cap when it shrank again.[60] To keep the cabin cool, Concorde used the fuel as a heat sink for the heat from the air conditioning.[61] The same method also cooled the hydraulics. During supersonic flight the surfaces forward from the cockpit became heated, and a visor was used to deflect much of this heat from directly reaching the cockpit.[62]

Concorde had livery restrictions; the majority of the surface had to be covered with a highly reflective white paint to avoid overheating the aluminium structure due to heating effects from supersonic flight at Mach 2. In 1996, Air France briefly painted F-BTSD in a predominantly blue livery, with the exception of the wings, in a promotional deal with Pepsi.[63] In this paint scheme, Air France were advised to remain at Mach 2 for no more than 20 minutes at a time, but there was no restriction at speeds under Mach 1.7. F-BTSD was used because it was not scheduled for any long flights that required extended Mach 2 operations.[64]

Structural issues

Due to the high speeds at which Concorde travelled, large forces were applied to the aircraft's structure during banks and turns. This caused twisting and the distortion of the aircraft's structure. In addition there were concerns over maintaining precise control at supersonic speeds; both of these issues were resolved by active ratio changes between the inboard and outboard elevons, varying at differing speeds including supersonic. Only the innermost elevons, which are attached to the stiffest area of the wings, were active at high speed.[65] Additionally, the narrow fuselage meant that the aircraft flexed.[36] This was visible from the rear passengers' viewpoints.[66]

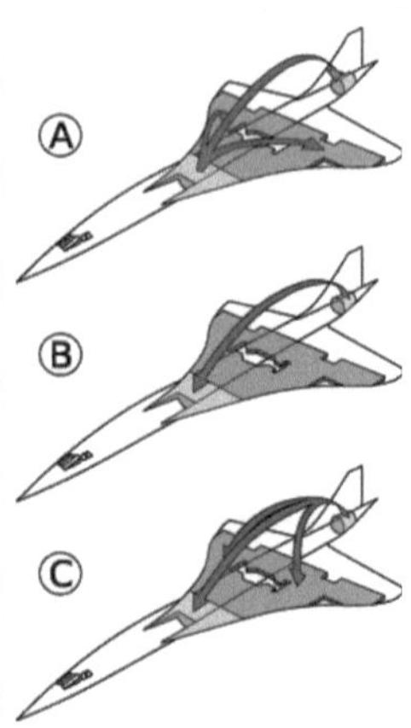

Fuel pitch trim

When any aircraft passes the critical mach of that particular airframe, the centre of pressure shifts rearwards. This causes a pitch down force on the aircraft if the centre of mass remains where it was. The engineers designed the wings in a specific manner to reduce this shift, but there was still a shift of about 2 metres. This could have been countered by the use of trim controls, but at such high speeds this would have caused a dramatic increase in the drag on the aircraft. Instead, the distribution of fuel along the aircraft was shifted during acceleration and deceleration to move the centre of mass, effectively acting as an auxiliary trim control.[67]

Range

In order to fly non-stop across the Atlantic Ocean, Concorde was developed to have the greatest supersonic range of any aircraft. This was achieved by a combination of engines which were highly efficient at supersonic speeds,[68] [36] a slender fuselage with high fineness ratio, and a complex wing shape for a high lift to drag ratio. This also required carrying only a modest payload and a high fuel capacity, and the aircraft was trimmed with precision to avoid unnecessary drag.[7] [67]

Nevertheless, soon after Concorde began flying, a Concorde "B" model was designed with slightly larger fuel capacity and slightly larger wings with leading edge slats to improve aerodynamic performance at all speeds, with the objective of expanding the range to reach markets in new regions.[69] It featured more powerful engines with sound deadening and without the fuel-hungry and noisy reheat. It was speculated that it was reasonably possible to create an engine with up to 25% gain in efficiency over the Rolls-Royce/Snecma Olympus 593.[70] This would have given 500 mi (805 km) additional range and a greater payload, making new commercial routes possible. This was cancelled due in part to poor sales of Concorde, but also to the rising cost of aviation fuel in the 1970s.[71]

Cabin pressurisation and radiation concerns

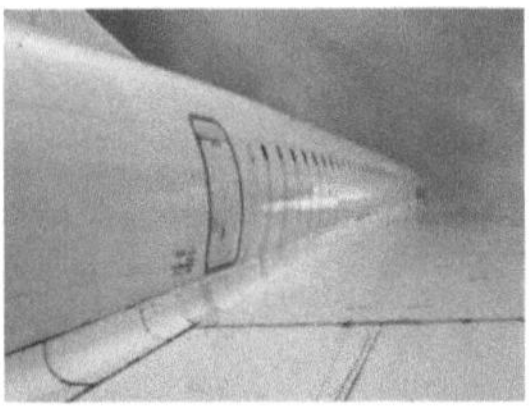

External view of Concorde's fuselage

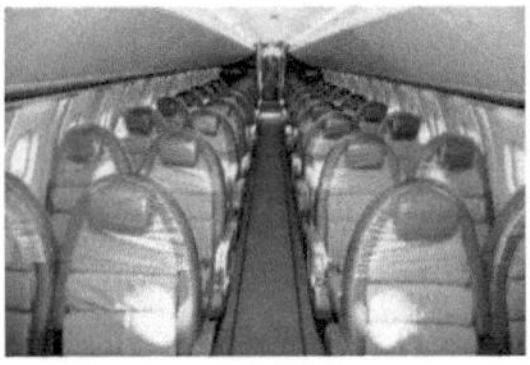

British Airways Concorde interior. The narrow fuselage permitted only 4 seats across the aircraft with limited headroom

The high altitude at which Concorde cruised meant passengers received almost twice the flux of extraterrestrial ionising radiation as those travelling on a conventional long-haul flight.[72] [73] Upon Concorde's introduction, it was speculated that this exposure during supersonic travels would increase the likelihood of skin cancer.[74] Due to the proportionally reduced flight time, the overall equivalent dose would normally be less than a conventional flight over the same distance.[75] Unusual solar activity might lead to an increase in incident radiation.[76] To prevent incidents of excessive radiation exposure, the flight deck had a radiometer and an instrument to measure the rate of decrease of radiation.[73] If the radiation level became too high, Concorde would descend below 47000 feet (m).

Airliner cabins were usually maintained at a pressure equivalent to 6,000–8,000 feet (1,800–2,400 m) elevation. Concorde's pressurisation was set to an altitude at the lower end of this range, 6000 feet (m).[77] Concorde's maximum cruising altitude was 60000 feet (m); subsonic airliners typically cruise below 40000 feet (m). Above 50000 feet (m), the lack of air pressure would give a "time of useful consciousness" in even a conditioned athlete of no more than 10–15 seconds.[78]

A sudden reduction in cabin pressure is hazardous to all passengers and crew.[79] At Concorde's altitude, the air density is very low; a breach of cabin integrity would result in a loss of pressure severe enough so that the plastic emergency oxygen masks installed on other passenger jets would not be effective, and passengers would quickly suffer from hypoxia despite quickly donning them. Concorde was equipped with smaller windows to reduce the rate of loss in the event of a breach,[80] a reserve air supply system to augment cabin air pressure, and a rapid descent procedure to bring the aircraft to a safe altitude. The FAA enforces minimum emergency descent rates for aircraft and made note of Concorde's higher operating altitude, concluding that the best response to a loss of pressure would be a rapid descent.[81] Continuous positive airway pressure would have delivered pressurised oxygen directly to the pilots through masks.[80]

Flight characteristics

Concorde performing a low-level flypast at an air show in August 1981

While commercial jets take eight hours to fly from New York to Paris, the average supersonic flight time on the transatlantic routes was just under 3.5 hours. Concorde had a maximum cruise altitude of **unknown operator: u','** metres (**unknown operator: u'strong'** ft) and an average cruise speed of Mach 2.02, about 1155 knots (2140 km/h or 1334 mph), more than twice the speed of conventional aircraft.[82]

With no other civil traffic operating at its cruising altitude of about 56000 ft (m), dedicated oceanic airways or "tracks" were used by Concorde to cross the Atlantic. Due to the nature of high altitude winds, these SST tracks were fixed in terms of their co-ordinates, unlike the North Atlantic Tracks at lower altitudes whose co-ordinates alter daily according to forecast weather patterns.[83] Concorde would also be cleared in a 15000-foot (m) block, allowing for a slow climb from 45,000 to 60000 ft (m) during the oceanic crossing as the fuel load gradually decreased.[84] In regular service, Concorde employed an efficient *cruise-climb* flight profile following take-off.[85]

The delta-shaped wings forced Concorde to attain a higher angle of attack at low speeds than conventional aircraft, but it allowed the formation of large low pressure vortices over the entire upper wing surface, maintaining lift.[86] The normal landing speed was 170 miles per hour (274 km/h).[87] Because of this high angle, during a landing approach Concorde was on the "back side" of the drag force curve, where raising the nose would increase the sink rate; the aircraft was thus largely flown on the throttle and was fitted with an autothrottle to reduce the pilot's workload.[88]

Brakes and undercarriage

Concorde rear undercarriage

Tail bumper of Concorde G-BOAG at the Museum of Flight in Seattle

Because of the way Concorde's delta-wing generated lift, the undercarriage had to be unusually strong. At rotation, Concorde would rise to a high angle of attack, about 18 degrees. Prior to rotation the wing generated almost no lift, unlike typical aircraft wings. Combined with the high airspeed at rotation (199 knots indicated airspeed), this increased the stresses on the rear undercarriage in a way that was initially unexpected during the development and required a major redesign.[89] Due to the high angle needed at rotation, a small set of wheels were added aft to prevent tailstrikes. The rear main undercarriage units swing towards each other to be stowed but due to their great height also need to retract telescopically before swinging in order to clear each other when stowed.[90]

Additionally, due to the high average takeoff speed of 250 miles per hour (km/h), Concorde needed upgraded brakes. Like most airliners, Concorde has anti-skid braking – a system which prevents the tyres from losing traction when the brakes are applied for greater control during roll-out. The brakes, developed by Dunlop, were the first carbon-based brakes used on an airliner.[91] The use of carbon over equivalent steel brakes provided a weight-saving of 12000 lb (5443 kg).[92]

Droop nose

Concorde's drooping nose, developed by Marshall Aerospace,[93] enabled the aircraft to switch between being streamlined to reduce drag and achieve optimum aerodynamic efficiency, and not obstructing the pilot's view during taxi, takeoff, and landing operations. Due to the high angle of attack the long pointed nose obstructed the view and necessitated the capability to droop. The droop nose was accompanied by a moving visor that retracted into the nose prior to being lowered. When the nose was raised to horizontal, the visor would raise in front of the cockpit windscreen for aerodynamic streamlining.[93]

Concorde landing at Farnborough in September 1974.

A controller in the cockpit allowed the visor to be retracted and the nose to be lowered to 5° below the standard horizontal position for taxiing and takeoff. Following takeoff and after clearing the airport, the nose and visor were raised. Prior to landing, the visor was again retracted and the nose lowered to 12.5° below horizontal for maximum visibility. Upon landing the nose was raised to the five-degree position to avoid the possibility of damage.[93]

The Federal Aviation Administration had objected to the restrictive visibility of the visor used on the first two prototype Concordes and thus requiring alteration before the FAA would permit Concorde to serve US airports; this led to the redesigned visor used on the production and the four pre-production aircraft (101, 102, 201, and 202).[94] The nose window and visor glass needed to endure temperatures in excess of 100°C at supersonic flight were developed by Triplex.[95]

Operational history

Scheduled flights

Scheduled flights began on 21 January 1976 on the London–Bahrain and Paris–Rio (via Dakar) routes,[96] with BA flights using the "*Speedbird Concorde*" call sign to notify air traffic control of the aircraft's unique abilities and restrictions, but the French using their normal callsigns.[97] The Paris-Caracas route (via Azores) began on 10 April. The US Congress had just banned Concorde landings in the US, mainly due to citizen protest over sonic booms, preventing launch on the coveted transatlantic routes. The US Secretary of Transportation, William Coleman, gave permission for Concorde service to Washington Dulles International Airport, and Air France and British Airways simultaneously began service to Dulles on 24 May 1976.[98]

The official handover ceremony to British Airways of its first Concorde occurred on 15 January 1976 at Heathrow Airport

When the US ban on JFK Concorde operations was lifted in February 1977, New York banned Concorde locally. The ban came to an end on 17 October 1977 when the Supreme Court of the United States declined to overturn a lower court's ruling rejecting efforts by the Port Authority and a grass-roots campaign led by Carol Berman to continue the ban.[99] In spite of complaints about noise, the noise report noted that Air Force One, at the time a Boeing VC-137, was louder than Concorde at subsonic speeds and during takeoff and landing.[100] Scheduled service from Paris and London to New York's John F. Kennedy Airport began on 22 November 1977.[101]

Concorde in Singapore Airlines livery at Heathrow in 1979

In 1977, British Airways and Singapore Airlines shared a Concorde for flights between London and Singapore International Airport at Paya Lebar via Bahrain. The aircraft, BA's Concorde G-BOAD, was painted in Singapore Airlines livery on the port side and British Airways livery on the starboard side.[102] [103] The service was discontinued after three return flights because of noise complaints from the Malaysian government;[104] it could only be reinstated on a new route bypassing Malaysian airspace in 1979. A dispute with India prevented Concorde from reaching supersonic speeds in Indian airspace, so the route was eventually declared not viable and discontinued in 1980.[105]

Air France Concorde in 1977

During the Mexican oil boom, Air France flew Concorde twice weekly to Mexico City's Benito Juárez International Airport via Washington, DC, or New York City, from September 1978 to November 1982.[106] [107] The worldwide economic crisis during that period resulted in this route's cancellation; the last flights were almost empty. The routing between Washington or New York and Mexico City included a deceleration, from Mach 2.02 to Mach 0.95, to cross Florida subsonically and avoid creating a sonic boom over the state; Concorde then re-accelerated back to high speed while crossing the Gulf of Mexico. On 1 April 1989, on an around-the-world luxury tour charter, British Airways implemented changes to this routing that allowed G-BOAF to maintain Mach 2.02 by passing around Florida to the east and south. Periodically Concorde visited the region on similar chartered flights to Mexico City and Acapulco.[108]

From 1978 to 1980, Braniff International Airways leased 10 Concordes, five each from Air France and British Airways.[109] These were used on subsonic flights between Dallas-Fort Worth and Washington Dulles International

Airport, flown by Braniff flight crews.[110] Air France and British Airways crews then took over for the continuing supersonic flights to London and Paris.[111] The aircraft were registered in both the United States and their home countries; the European registration was covered while being operated by Braniff, retaining full AF/BA liveries. The flights were not profitable and typically less than 50% booked, forcing Braniff to end its tenure as the only US Concorde operator in May 1980.[112] [113]

British Caledonian interest

Following the launch of British Airways' Concorde services, Britain's other major airline, British Caledonian (BCal), set up a task force headed by Gordon Davidson, BA's former Concorde director, to investigate the possibility of their own Concorde operations.[114] [115] [116] This was seen as particularly viable for the airline's long-haul network as there were two unsold aircraft then available for purchase.[117] [118] [119]

One important reason for BCal's interest in Concorde was that the British Government's 1976 aviation policy review had opened the possibility of BA setting up supersonic services in competition with BCal's established sphere of influence. To counteract this potential threat, BCal considered their own independent Concorde plans, as well as a partnership with BA.[120] [121] BCal were considered most likely to have set up a Concorde service on the Gatwick–Lagos route, a major source of revenue and profits within BCal's scheduled route network;[122] [123] BCal's Concorde task force did assess the viability of a daily supersonic service complementing the existing subsonic widebody service on this route.[118] [121] [124]

BCal entered into a bid to acquire at least one Concorde.[117] [119] [124] However, BCal eventually arranged for two aircraft to be leased from BA and Aérospatiale respectively, to be maintained by either BA or Air France. BCal's envisaged two-Concorde fleet would have required a high level of aircraft utilisation to be cost-effective; therefore, BCal had decided to operate the second aircraft on a supersonic service between Gatwick and Atlanta, with a stopover at either Gander or Halifax.[118] Consideration was given to services to Houston and various points on its South American network at a later stage.[124] [125] Both supersonic services were to be launched at some point during 1980; however, steeply rising oil prices caused by the 1979 energy crisis led to BCal shelving their supersonic ambitions.[121]

British Airways buys its Concordes outright

By around 1981 in the UK, the future for Concorde looked bleak. The British government had lost money operating Concorde every year, and moves were afoot to cancel the service entirely. A cost projection came back with greatly reduced metallurgical testing costs because the test rig for the wings had built up enough data to last for 30 years and could be shut down. Despite this, the government was not keen to continue. In 1983, BA's managing director, Sir John King, convinced the government to sell the aircraft outright to British Airways for £16.5 million plus the first year's profits.[126] [127]

An Air France Concorde at John F. Kennedy International Airport in 1987

King recognised that, in Concorde, BA had a premier product that was underpriced. Market research had revealed that many customers thought Concorde was more expensive than it actually was; thus ticket prices were progressively raised to match these perceptions.[36] It is reported that British Airways then ran Concorde at a profit, unlike their French counterpart.[128] [129]

Between 1984 and 1991, British Airways flew a thrice-weekly Concorde service between London and Miami, stopping at Washington Dulles International Airport.[130] [131] Until 2003, Air France and British Airways continued to operate the New York services daily. Concorde routinely flew to Grantley Adams International Airport, Barbados, during the winter holiday season.[132]

Prior to the Air France Paris crash, several UK and French tour operators operated charter flights to European destinations on a regular basis;[133] [134] the charter business was viewed as lucrative by British Airways and Air France.[135]

Concorde Flight 4590 crash

On 25 July 2000, Air France Flight 4590, registration F-BTSC, crashed in Gonesse, France after departing from Paris Charles de Gaulle en route to John F. Kennedy International Airport in New York City, killing all 100 passengers and nine crew members on board the flight, and four people on the ground. It was the only fatal accident involving Concorde.

According to the official investigation conducted by the *Bureau d'Enquêtes et d'Analyses pour la Sécurité de l'Aviation Civile* (BEA), the crash was caused by a titanium strip that fell from a Continental Airlines DC-10 that had taken off minutes earlier. This metal fragment punctured a tyre on Concorde's left main wheel bogie during takeoff. The tyre exploded, a piece of rubber hit the fuel tank, and while the fuel tank was not punctured, the impact caused a shock-wave which caused one of the fuel valves in the wing to burst open. This caused a major fuel leak from the tank, which then ignited due to sparking electrical landing gear wiring severed by another piece of the same tyre. The crew shut down engine number 2 in response to a fire warning, and with engine number 1 surging and producing little power, the aircraft was unable to gain height or speed. The aircraft entered a rapid pitch-up then a violent descent, rolling left and crashing tail-low into the Hôtelissimo Les Relais Bleus Hotel in Gonesse.[136] On 6 December 2010, Continental Airlines and John Taylor, one of their mechanics, were found guilty of involuntary manslaughter,[137] but on 30 November 2012 a French court overturned the conviction, saying mistakes by Continental and Taylor did not make them criminally responsible.[138]

Prior to the accident, Concorde had been arguably the safest operational passenger airliner in the world in terms of passenger deaths-per-kilometres travelled with zero, but with a history of tyre explosions 60 times higher than subsonic jets.[139] Safety improvements were made in the wake of the crash, including more secure electrical controls, Kevlar lining to the fuel tanks and specially developed burst-resistant tyres.[140]

The first flight after the modifications departed from London Heathrow on 17 July 2001, piloted by BA Chief Concorde Pilot Mike Bannister. During the 3-hour 20-minute flight over the mid-Atlantic towards Iceland, Bannister attained Mach 2.02 and 60000 ft (m) before returning to RAF Brize Norton. The test flight, intended to resemble the London–New York route, was declared a success and was watched on live TV, and by crowds on the ground at both locations.[141] The first flight with passengers after the accident took place on 11 September 2001, which landed shortly before the World Trade Center attacks in the United States. This was not a revenue flight, as all the passengers were BA employees.[142]

Normal commercial operations resumed on 7 November 2001 by BA and AF (aircraft G-BOAE and F-BTSD), with service to New York JFK, where passengers were welcomed by then mayor Rudy Giuliani.[143] [144]

Retirement

Concorde G-BOAB in storage at London Heathrow Airport. This aircraft flew for 22,296 hours between its first flight in 1976 and its final flight in 2000.

On 10 April 2003, Air France and British Airways simultaneously announced that they would retire Concorde later that year.[145] They cited low passenger numbers following the 25 July 2000 crash, the slump in air travel following 11 September 2001, and rising maintenance costs. Although Concorde was technologically advanced when introduced in the 1970s, 30 years later its analogue cockpit was dated. There had been little commercial pressure to upgrade Concorde due to a lack of competing aircraft, unlike other airliners of the same era such as the Boeing 747.[146] By its retirement, it was the last aircraft in British Airways' fleet that had a flight engineer; other aircraft, such as the modernised 747-400, had eliminated the role.[147]

On the same day, Sir Richard Branson offered to buy British Airways' Concorde fleet at their "original price of £1" for service with Virgin Atlantic Airways. Branson claimed this to be the same token price that British Airways had paid the British government; however, BA denied this and refused the offer.[148] Branson wrote in *The Economist* (23 October 2003) that his final offer was "over £5 million" and that he had intended to operate the fleet "for many years to come".[149] The chances for keeping Concorde in service were stifled by Airbus's lack of support for continued maintenance.[150] [151] [152]

It has been suggested that Concorde was not withdrawn for the reasons usually given but that it became apparent during the grounding of Concorde that the airlines could make more profit carrying first class passengers subsonically.[153] Rob Lewis suggested that the Air France retirement of its Concorde fleet was the result of a conspiracy between Air France Chairman Jean-Cyril Spinetta and Airbus CEO Noel Forgeard, and stemmed as much from a fear of being found criminally liable under French law for future Concorde accidents as from simple economics.[154] A lack of commitment to Concorde from Director of Engineering Alan MacDonald was cited as having undermined BA's resolve to continue operating Concorde.[155]

Promotional efforts in 1997 included a British Airways lottery to fly to New York held for 190 tickets valued at £5,400 each, to be offered up at £10. Callers had to ring a special hotline competing with up to 20 million people. The offer marked the 10th anniversary of BA's move into the private sector. Among the winners were 15 Scottish residents, including Joe Traynor and wife Margaret. Other winners included a health service worker from Glasgow who only asked for one ticket saying he couldn't think of anyone to take with him. Health Service Worker a Murali Murugasu, 28, said: "I didn't have time to tell anyone or ring them up so I will have to spend the time on my own. I am doing it just for the sheer experience of going on Concorde. At the time, British Airways were baffled why so many Scots got lucky. A spokeswoman said: "It was a good proportion, considering it was a worldwide promotion.[156]

Air France

Air France made its final commercial Concorde landing in the United States in New York City from Paris on 30 May 2003.[157] [158] Air France's final Concorde flight took place on 27 June 2003 when F-BVFC retired to Toulouse.[159]

An auction of Concorde parts and memorabilia for Air France was held at Christie's in Paris on 15 November 2003; 1,300 people attended, and several lots exceeded their predicted values.[160] French Concorde F-BVFC was retired to Toulouse and kept functional for a short time after the end of service, in case taxi runs were required in support of the French judicial enquiry into the 2000 crash.[161] The aircraft is now fully retired and no longer functional.[162]

Air France Concorde at Paris-Charles de Gaulle Airport

French Concorde F-BTSD has been retired to the "Musée de l'Air et de l'Espace" at Le Bourget (near Paris) and, unlike the other museum Concordes, a few of the systems are being kept functional. For instance, the famous "droop nose" can still be lowered and raised. This led to rumours that they could be prepared for future flights for special occasions.[163]

French Concorde F-BVFB currently rests at the Auto & Technik Museum Sinsheim at Sinsheim, Germany, after its last flight from Paris to Baden-Baden, followed by a spectacular transport to Sinsheim via barge and road. The museum also has a Tu-144 on display – this is the only place where both supersonic airliners can be seen together.[164]

British Airways

British Airways conducted a North American farewell tour in October 2003. G-BOAG visited Toronto Pearson International Airport on 1 October, after which it flew to New York's John F. Kennedy International Airport.[165] G-BOAD visited Boston's Logan International Airport on 8 October, and G-BOAG visited Washington Dulles International Airport on 14 October.[166] It has been claimed that G-BOAD's flight from London Heathrow to Boston set a transatlantic flight record of 3 hours, 5 minutes, 34 seconds.[167] However the fastest transatlantic flight was from Heathrow to New York JFK airport on 7 February 1996, taking only 2 hours, 52 minutes, 59 seconds.[168] [169]

Mike Bannister (left) in the cockpit of BA002

In a week of farewell flights around the United Kingdom, Concorde visited Birmingham on 20 October, Belfast on 21 October, Manchester on 22 October, Cardiff on 23 October, and Edinburgh on 24 October. Each day the aircraft made a return flight out and back into Heathrow to the cities, often overflying them at low altitude.[170] [171] [172] On 22 October, both Concorde flight BA9021C, a special from Manchester, and BA002 from New York landed simultaneously on both of Heathrow's runways. On 23 October 2003, the Queen consented to the illumination of Windsor Castle, an honour reserved for state events and visiting dignitaries, as Concorde's last west-bound commercial flight departed London.[173]

Concorde G-BOAC in her hangar at Manchester Airport Aviation Viewing Park

British Airways retired its Concorde fleet on 24 October.[174] G-BOAG left New York to a fanfare similar to that given for Air France's F-BTSD, while two more made round trips, G-BOAF over the Bay of Biscay, carrying VIP guests including former Concorde pilots, and G-BOAE to Edinburgh. The three aircraft then circled over London, having received special permission to fly at low altitude, before landing in sequence at Heathrow. The captain of the New York to London flight was Mike Bannister.[175]

The final flight of a Concorde in the US occurred on 5 November 2003 when G-BOAG flew from New York's Kennedy Airport to Seattle's Boeing Field to join the Museum of Flight's permanent collection. The plane was piloted by Mike Bannister and Les Broadie who claimed a flight time of three hours, 55 minutes and 12 seconds, a record between the two cities.[176] The museum had been pursuing a Concorde for their collection since 1984.[177] The final flight of a Concorde world-wide took place on 26 November 2003 with a landing at Filton, Bristol, UK.[178]

All of BA's Concorde fleet have been grounded, drained of hydraulic fluid and their airworthiness certificates withdrawn. Jock Lowe, ex-chief Concorde pilot and manager of the fleet estimated in 2004 that it would cost £10–15 million to make G-BOAF airworthy again.[163] BA maintain ownership and have stated that they will not fly again due to a lack of support from Airbus.[179] On 1 December 2003, Bonhams held an auction of British Airways' Concorde artifacts, including a nose cone, at Kensington Olympia in London.[180] [181] Proceeds of around £750,000 were raised, with the majority going to charity. G-BOAD is currently on display at the Intrepid Sea, Air & Space Museum in New York.[169] In 2007, BA announced that the advertising spot at Heathrow where a 40% scale model of Concorde was located would not be retained; the model is now on display at the Brooklands Museum.[182]

Restoration

Although only used for spares after being retired from test flying and trials work in 1981, Concorde G-BBDG was dismantled and transported by road from Filton then restored from essentially a shell at the Brooklands Museum in Surrey.[183]

One of the youngest Concordes (F-BTSD) is on display at Le Bourget Air and Space Museum in Paris. In February 2010, it was announced that the museum and a group of volunteer Air France technicians intend to restore F-BTSD so it can taxi under its own power.[184] On 29 May 2010, it was reported that a group comprising the British Save Concorde Group and the French Olympus 593 had begun work on inspecting the engines of a Concorde at Le Bourget Air and Space Museum, with the intent to restore the plane to be able to fly again in demonstrations and air shows.[185]

Comparable aircraft

The only supersonic airliner in direct competition with Concorde was the Soviet Tupolev Tu-144, nicknamed "Concordski" by Western Europeans for its outward similarity to Concorde.[186] It had been alleged that Soviet espionage efforts had resulted in the theft of Concorde blueprints, ostensibly to assist in the design of the Tu-144.[187] As a result of a rushed development programme, the first Tu-144 prototype was substantially different from the preproduction machines, but both were cruder than Concorde. The Tu-144*S* had a significantly shorter range than Concorde, due to its low-bypass turbofan engines.[188] The aircraft had poor control at low speeds because of a simpler supersonic wing design; in addition the Tu-144 required parachutes to land while Concorde used anti-lock brakes.[189] The Tu-144 had two crashes, one at the 1973 Paris Air Show,[190] [191] and another during a pre-delivery test flight in May 1978.[192] [193]

Later production Tu-144 versions were more refined and competitive. They had retractable canards for better low-speed control, turbojet engines providing nearly the fuel efficiency and range of Concorde[194] and a top speed of Mach 2.35. Passenger service commenced in November 1977, but after the 1978 crash the aircraft was taken out of service. The aircraft had an inherently unsafe structural design as a consequence of an automated production method chosen in order to simplify and speed up manufacturing.[195]

The American designs, the Boeing 2707 and the Lockheed L-2000 were to have been larger, with seating for up to 300 people.[196] [197] Running a few years behind Concorde, the Boeing 2707 was redesigned to a cropped delta layout; the extra cost of these changes helped to kill the project.[198] The operation of US military aircraft such as the XB-70 Valkyrie and B-58 Hustler had shown that sonic booms were quite capable of reaching the ground,[199] and the experience from the Oklahoma City sonic boom tests led to the same environmental concerns that hindered the commercial success of Concorde. The American government cancelled the project in 1971, after having spent more than $1 billion.[200]

The only other large supersonic aircraft comparable to Concorde are strategic bombers, principally the Russian Tupolev Tu-22, Tu-22M and Tu-160 and the American B-1 Lancer.[201]

Impact

Environmental

Prior to Concorde's flight trials, the developments made by the civil aviation industry were largely accepted by governments and their respective electorates. Opposition to Concorde's noise, particularly on the east coast of the United States,[202] [203] forged a new political agenda on both sides of the Atlantic, with scientists and technology experts across a multitude of industries beginning to take the environmental and social impact more seriously.[204] [205] Although Concorde led directly to the introduction of a general noise abatement programme for aircraft flying out of John F. Kennedy Airport, many found that Concorde was quieter than expected,[36] partly due to the pilots temporarily throttling back their engines to reduce noise during overflight of residential areas.[206] Even before the launch of revenue earning services, it had been noted that Concorde was quieter than several aircraft already commonly in service at that time.[207]

Concorde produced nitrogen oxides in its exhaust, which, despite complicated chemical interactions with other ozone-depleting chemicals, are understood to result in degradation to the ozone layer at the stratospheric altitudes it cruised.[208] It has been pointed out that other, lower-flying, airliners produce ozone during their flights in the troposphere, but vertical transit of gases between the layers is restricted. The small fleet operated meant overall ozone-layer degradation caused by Concorde was negligible.[208] David W. Fahey, of the National Oceanic and Atmospheric Administration, found that to produce a drop in stratospheric Ozone of 1 to 2% would require a fleet of 500 supersonic aircraft to be operated. Dr. Fahey stated that this would not be a limiting factor for further Supersonic Transport development.[209]

Concorde's technical leap forward boosted the public's understanding of conflicts between technology and the environment as well as awareness of the complex decision analysis processes that surround such conflicts.[210] In France, the use of acoustic fencing alongside TGV tracks might not have been achieved without the 1970s controversy over aircraft noise.[211] In the UK, the CPRE has issued tranquillity maps since 1990.[212]

Public perception

Parade flight at Queen's Golden Jubilee in June 2002

HM The Queen and HRH The Duke of Edinburgh disembark Concorde in 1991.

Concorde was normally perceived as a privilege of the rich, but special circular or one-way (with return by other flight or ship) charter flights were arranged to bring a trip within the means of moderately well-off enthusiasts.[213]

The aircraft was usually referred to by the British as simply "Concorde".[214] In France it was known as "le Concorde" due to "le", the definite article,[215] used in French grammar to introduce the name of a ship or aircraft,[216] and the capital being used to distinguish a proper name from a common noun of the same spelling.[215] [217] In French, the common noun *concorde* means "agreement, harmony, or peace". [218] Concorde's pilots and British Airways in official publications often refer to Concorde both in the singular and plural as "she" or "her".[219] [220] [221]

As a symbol of national pride, an example from the BA fleet made occasional flypasts at selected Royal events, major air shows and other special occasions, sometimes in formation with the Red Arrows.[222] [223] On the final day of commercial service, public interest was so great that grandstands were erected at Heathrow Airport. Significant numbers of people attended the final landings; the event received widespread media coverage.[224]

In 2006, 37 years after her first test flight, Concorde was announced the winner of the Great British Design Quest organised by the BBC and the Design Museum. A total of 212,000 votes were cast with Concorde beating design icons such as the Mini, mini skirt, Jaguar E-type, Tube map and the Supermarine Spitfire.[5]

Records

The fastest transatlantic airliner flight was from New York JFK to London Heathrow on 7 February 1996 by British Airways' G-BOAD in 2 hours, 52 minutes, 59 seconds from takeoff to touchdown.[168] Concorde also set other records, including the official FAI "Westbound Around the World" and "Eastbound Around the World" world air speed records.[225] On 12–13 October 1992, in commemoration of the 500th anniversary of Columbus' first New World landing, Concorde Spirit Tours (USA) chartered Air France Concorde F-BTSD and circumnavigated the world in 32 hours 49 minutes and 3 seconds, from Lisbon, Portugal, including six refuelling stops at Santo Domingo, Acapulco, Honolulu, Guam, Bangkok, and Bahrain.[226]

The eastbound record was set by the same Air France Concorde (F-BTSD) under charter to Concorde Spirit Tours in the USA on 15–16 August 1995. This promotional flight circumnavigated the world from New York/JFK International Airport in 31 hours 27 minutes 49 seconds, including six refuelling stops at Toulouse, Dubai, Bangkok, Andersen AFB in Guam, Honolulu, and Acapulco.[227] By its 30th flight anniversary on 2 March 1999 Concorde had clocked up 920,000 flight hours, with more than 600,000 supersonic, much more than all of the other supersonic aircraft in the Western world combined.[228]

On its way to the Museum of Flight in November 2003, G-BOAG set a New York City-to-Seattle speed record of 3 hours, 55 minutes, and 12 seconds.[229]

Replacements in development

The desire for a second-generation supersonic aircraft has remained within some elements of the aviation industry,[230] [231] and several concepts emerged quickly following the retirement of Concorde.

In November 2003, EADS—the parent company of Airbus—announced that it was considering working with Japanese companies to develop a larger, faster replacement for Concorde.[232] [233] In October 2005, JAXA, the Japan Aerospace eXploration Agency, undertook aerodynamic testing of a scale model of an airliner designed to carry 300 passengers at Mach 2 (working name *NEXST*). If pursued to commercial deployment, it would be expected to be in service around 2020–2025.[234] On 18 June 2011, the *Zero Emission High Speed Transport* or *ZEHST* concept aircraft was unveiled by EADS at the Paris Air Show.[235] The ZEHST, a hypersonic aircraft to be capable of 3,000 mph (4,800 km/h), is a result of the collaboration efforts between EADS and Japan.[236]

The British company Reaction Engines Limited, with 50% EU money, has been engaged in a research programme called *LAPCAT*, which examined a design for a hydrogen-fuelled plane carrying 300 passengers called the *A2*, potentially capable of flying at Mach 5+ non-stop from Brussels to Sydney in 4.6 hours.[237] The follow-on research effort, *LAPCAT II* began in 2008 and is to last four years.[238]

In May 2008, it was reported that Aerion Corporation had $3 billion of pre-order sales on its Aerion SBJ supersonic business jet.[239] In late 2010, the project continued with a testbed flight of a section of the wing.[240]

Supersonic Aerospace International's Quiet Supersonic Transport was a 12 passenger design from Lockheed Martin that was to cruise at Mach 1.6, and was to have created a sonic boom only 1% as strong as that generated by Concorde.[241]

Operators

- Air France
- British Airways
- Braniff International Airways (short term lease)[112]
- Singapore Airlines (short term wet lease)[103]

Specifications

Data from Wall Street Journal,[147] *The Concorde Story,*[242] *The International Directory of Civil Aircraft,*[38] *Richard Seamen aircraft museum*[243]

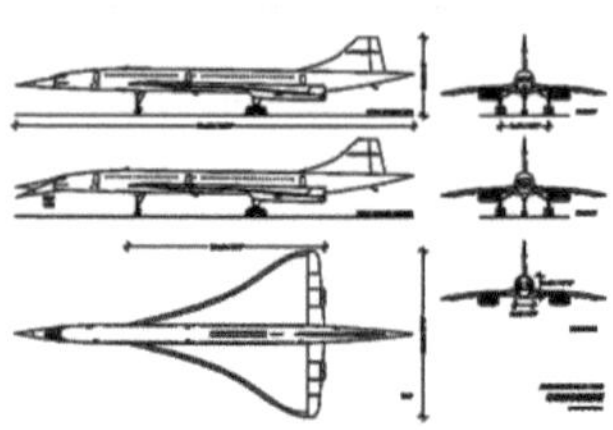

Other line drawings of Concorde

General characteristics

- **Crew:** 3 (2 Pilots and a flight engineer)
- **Capacity:** 92–120 passengers (128 in high-density layout)[244]
- **Length:** 202 ft 4 in (61.66 m)
- **Wingspan:** 84 ft 0 in (25.6 m)
- **Height:** 40 ft 0 in (12.2 m)
- **Fuselage internal length:** 129 ft 0 in (m)
- **Fuselage width:** maximum of 9 ft 5 in (m) external 8 ft 7 in (m) internal
- **Fuselage height:** maximum of 10 ft 10 in (m) external 6 ft 5 in (m) internal)
- **Wing area:** 3,856 ft^2 (358.25 m^2)

- **Empty weight:** 173,500 lb (78,700 kg)
- **Useful load:** 245,000 lb (111,130 kg)
- **Powerplant:** 4 × Rolls-Royce/SNECMA Olympus 593 Mk 610 afterburning turbojets
  - **Dry thrust:** 32,000 lbf (140 kN) each
  - **Thrust with afterburner:** 38,050 lbf (169 kN) each
- **Maximum fuel load:** 210940 lb (kg)
- **Maximum taxiing weight:** 412000 lb (kg)

Performance

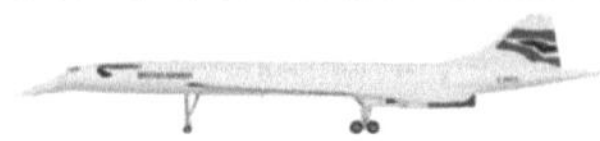
Concorde G-BOAC

- **Maximum speed:** Mach 2.04 (≈1,354 mph, 2,179 km/h) at cruise altitude
- **Cruise speed:** Mach 2.02 (≈1,340 mph, 2,158 km/h) at cruise altitude
- **Range:** 3,900 nmi (4,500 mi, 7,250 km)
- **Service ceiling:** 60,000 ft (18,300 m)
- **Rate of climb:** 5,000 ft/min (25.41 m/s)
- **lift-to-drag:** *Low speed*– 3.94, *Approach*– 4.35, *250 kn, 10,000 ft*– 9.27, *Mach 0.94*– 11.47, *Mach 2.04*– 7.14
- **Fuel consumption:** 46.85 lb/mi (13.2 kg/km) operating for maximum range
- **Thrust/weight:** 0.373
- **Maximum nose tip temperature:** 260 °F (°C)

See also

- BAC 221
- Barbara Harmer, the first qualified female Concorde pilot.
- North American XB-70 Valkyrie
- Richard Wiggs, Anti-Concorde campaigner

Related development

- Bristol Type 223
- Sud Aviation Super-Caravelle
- Fairey Delta

Related lists

- List of civil aircraft

References

Notes

[1] Towey 2007, p. 359.
[2] "Ageing luxury jet" (http://news.bbc.co.uk/1/hi/uk/850899.stm). *BBC News*, 25 July 2000. 25 July 2000. .
[3] Marston, Paul (16 August 2000). "Is this the end of the Concorde dream?" (http://www.telegraph.co.uk/news/uknews/1352927/Is-this-the-end-of-the-Concorde-dream.html). London: *The Daily Telegraph*, 16 August 2000. .
[4] http://news.bbc.co.uk/2/hi/uk_news/2934257.stm
[5] Jury, Louise (17 March 2006). "Concorde beats Tube map to become Britain's favourite design" (http://news.independent.co.uk/uk/this_britain/article351851.ece). London: *The Independent*, 17 March 2006. .
[6] "Early History." (http://www.concordesst.com/history/eh1.html#a) *concordesst.com*. Retrieved 8 September 2007.
[7] Maltby, R.L.. "The development of the slender delta concept" (http://md1.csa.com/partners/viewrecord.php?requester=gs&collection=TRD&recid=A6825061AH&q=Concorde+Delta&uid=788872723&setcookie=yes). *Aircraft Engineering* **40, 1968**. .
[8] "Aerospace: Pan Am's Concorde Retreat" (http://www.time.com/time/magazine/article/0,9171,903850,00.html). *Time*, 12 February 1973. 12 February 1973. .
[9] "Vertrag mit Luken" (http://www.spiegel.de/spiegel/print/d-46437753.html). Der Spiegel. 13 March 1967. . Retrieved 6 November 2012.

[10] Taylor 1965, p. 130.
[11] Winchester 2005, p. 134.
[12] Benn, Tony (17 October 2003). "Sonic booms and that 'e' on the end: Tony Benn remembers his role in getting Concorde off the ground" (http://www.guardian.co.uk/g2/story/0,3604,1064752,00.html). London: *The Guardian*, 17 October 2003. .
[13] McIntyre 1992, p. 20.
[14] Note this British convention is used throughout this article: "In depth: Farewell to Concorde" (http://news.bbc.co.uk/2/hi/in_depth/uk/2003/concorde_retirement/default.stm). *BBC News*, 15 August 2007. 15 August 2007. .
[15] "About Concorde – main page" (http://www.britishairways.com/concorde/aboutconcorde.html). *British Airways*. . Retrieved 11 January 2010.
[16] "Pilot Says Concorde Flight "Perfect"" (http://news.google.co.uk/newspapers?id=KXoyAAAAIBAJ&sjid=GLkFAAAAIBAJ&pg=6573,360672&dq=concorde&hl=en). *Montreal Gazette*,. 1969. . Retrieved 30 June 2011.
[17] "Concorde Tops Speed of Sound for 9 Minutes on a Test Flight" (http://select.nytimes.com/gst/abstract.html?res=F30C1EFC3454127B93C0A9178BD95F4D8685F9). *The New York Times*, 2 October 1969. 2 October 1969. .
[18] "1969: Concorde flies for the first time" (http://news.bbc.co.uk/onthisday/hi/dates/stories/march/2/newsid_2514000/2514535.stm). *BBC News*, 2 March 1969. 2 March 1969. . Retrieved 8 July 2007.
[19] Rohrbach, Edward (10 April 1969). "Concorde 002 Makes 1st Flight" (http://pqasb.pqarchiver.com/chicagotribune/access/584606362.html?dids=584606362:584606362&FMT=ABS&FMTS=ABS:AI&type=historic&date=Apr+10,+1969&author=&pub=Chicago+Tribune&desc=Concorde+002+Makes+1st+Flight&pqatl=google). *Chicago Tribune*, 10 April 1969. . Retrieved 30 June 2011.
[20] "Concorde 001 Makes Its First Atlantic Crossing" (http://pqasb.pqarchiver.com/chicagotribune/access/597531692.html?dids=597531692:597531692&FMT=CITE&FMTS=CITE:AI&type=historic&date=Sep+05,+1971&author=&pub=Chicago+Tribune&desc=Concorde+001+Makes+Its+First+Atlantic+Crossing&pqatl=google). *Chicago Tribune*, 5 September 1971. 5 September 1971. . Retrieved 30 June 2011.
[21] "Anglo-French Concorde Lands in Brazil to begin Week of Demonstration Flights" (http://news.google.co.uk/newspapers?id=VCY0AAAAIBAJ&sjid=2uAIAAAAIBAJ&pg=2521,1962837&dq=concorde&hl=en). Bangor Daily News. 7 September 1971. .
[22] Stern, Michael (3 June 1972). "Concorde Prototype Begins 10-Nation Tour; Britain Shows Optimism For Supersonic Aircraft" (http://select.nytimes.com/gst/abstract.html?res=F70D10F73B591A7493C1A9178DD85F468785F9). *The New York Times*, 3 June 1972. . Retrieved 30 June 2011.
[23] Witkin, Richard (21 September 1973). "A Supersonic Concorde Lands in Texas" (http://select.nytimes.com/gst/abstract.html?res=FA0E16FB3A54137A93C3AB1782D85F478785F9). *The New York Times*, 21 September 1973. . Retrieved 30 June 2011.
[24] "Concordes limited to 16". *Virgin Islands Daily News*, 5 June 1976.
[25] "Payments for Concorde" (http://www.britishairways.com/concorde/faq.html#6). *British Airways*. . Retrieved 2 December 2009.
[26] Lewis, Anthony (12 February 1973). "Britain and France have wasted billions on the Concorde" (http://select.nytimes.com/gst/abstract.html?res=F50D16F93C54137A93C0A81789D85F478785F9). *The New York Times*, 12 February 1973. .
[27] "Malaysia lifting ban on the use Of its Airspace by the Concorde" (http://select.nytimes.com/gst/abstract.html?res=F70813F83F5511728DDDAE0994DA415B888BF1D3). *The New York Times*, 17 December 1978. 17 December 1978. . Retrieved 30 June 2011.
[28] "News from around the world" (http://news.google.co.uk/newspapers?id=xTMsAAAAIBAJ&sjid=48sEAAAAIBAJ&pg=6855,2298265&dq=india+concorde&hl=en). *Herald-Journal*, 13 January 1978. . Retrieved 30 June 2011.
[29] Ross 1978, p. 46.
[30] Marston, Paul (16 August 2000). "Is this the end of the Concorde dream?" (http://www.telegraph.co.uk/news/uknews/1352927/Is-this-the-end-of-the-Concorde-dream.html). London: *Daily Telegraph*, 16 August 2000. .
[31] Ross 1978, pp. 47–49.
[32] Ross 1978, p. 49.
[33] Gunn 2010, p. 45.
[34] Favre, C. (1996). Mark B. Tischler. ed. *Advances in aircraft flight control* (http://books.google.com/?id=DbUhEnlI3OkC&pg=PA211&dq=wire+Concorde#v=onepage&q=wire Concorde&f=false). CRC Press. p. 219. ISBN 0-7484-0479-1. . Retrieved 30 June 2011.
[35] Masefield, Peter. "Obituary: Sir Archibald Russell". UK: *The Independent*, 1 July 1995.
[36] "NOVA transcript: Supersonic Dream" (http://www.pbs.org/wgbh/nova/transcripts/3203_concorde.html). *PBS*, 18 January 2005. .
[37] "Rolls-Royce Snecma Olympus" (http://www.janes.com/transport/news/jae/jae000725_1_n.shtml). Janes. 25 July 2000. .
[38] Frawley 2003, p. 14.
[39] "Unified propulsion 8" (http://ocw.mit.edu/ans7870/16/16.unified/propulsionS04/UnifiedPropulsion8/UnifiedPropulsion8.htm). MIT. . Retrieved 8 December 2010.
[40] "*Concorde* – Choice of a light alloy for the construction of the first supersonic commercial aircraft" (http://md1.csa.com/partners/viewrecord.php?requester=gs&collection=TRD&recid=A6424111AH&q=Concorde+aluminium+construction&uid=789267644&setcookie=yes). *Revue De L'Aluminium* (316, March 1964): 111–119. .
[41] Wolfe, B.S. (1967). "The Concorde Automatic Flight Control System: A description of the automatic flight control system for the Anglo/French SST and its development to date" (http://www.emeraldinsight.com/journals.htm?articleid=1681566&show=pdf). *Aircraft Engineering and Aerospace Technology* (MCB UP) **39** (5, 1967): 40. doi:10.1108/eb034268. ISSN 0002-2667. .

[42] Schefer, L.J.. "Concorde has designed-in reliability" (http://md1.csa.com/partners/viewrecord.php?requester=gs&collection=TRD&recid=A7716749AH&q=Concorde+hydraulic&uid=788858323&setcookie=yes). *Hydraulics and Pneumatics* **29, 1976**: 51–55. .

[43] Owen 2001, p. 101.

[44] "Aircraft Stopping Systems" (http://www.emeraldinsight.com/Insight/viewContentItem.do?contentType=Article&hdAction=lnkpdf&contentId=1682576). *Aircraft Engineering and Aerospace Technology* (MCB UP) **47** (10, 1975): 18. 1975. doi:10.1108/eb035278. ISSN 0002-2667. .

[45] Turner, H.G. (1971). "Fuel Management for Concorde: A brief account of the fuel system and the fuel pumps developed for the aircraft" (http://www.emeraldinsight.com/Insight/viewContentItem.do?contentType=Article&contentId=1682044). *Aircraft Engineering and Aerospace Technology* (MCB Ltd) **43** (3, 1971): 36–39. doi:10.1108/eb034738. ISSN 0002-2667. .

[46] "British Contribution to Concord Production in France" (http://www.emeraldinsight.com/Insight/viewContentItem.do?contentType=Article&hdAction=lnkpdf&contentId=1681209). *Aircraft Engineering and Aerospace Technology* (MCB Ltd) **36** (8, 1964): 232–237. 1964. doi:10.1108/eb033911. ISBN 0002-2667. .

[47] Owen 2001, p. 206.

[48] Birtles, Philip. *Concorde*, pp. 62–63. Vergennes, Vermont: Plymouth Press, 2000. ISBN 1-882663-44-6.

[49] "Rolls Royce Olympus history." (http://www.wingweb.co.uk/engines/Olympus593_MK610.html) *wingweb.co.uk*. Retrieved 15 January 2010.

[50] Ganley, G.; G. Laviec (1989). *The Rolls Royce/SNECMA Olympus 593 engine operational experience and the lessons learned* (http://md1.csa.com/partners/viewrecord.php?requester=gs&collection=TRD&recid=A9110958AH&q=Rolls-Royce/Snecma+Olympus+593&uid=788858323&setcookie=yes). *European Symposium on the Future of High Speed Air Transport*. pp. 73–80. . Retrieved 30 June 2011.

[51] Lynam, Joe (19 July 2006). "Are the skies turning green?" (http://news.bbc.co.uk/1/hi/business/5195964.stm). BBC News. .

[52] Ganley, G. A. (September 1991). *Concorde Propulsion—Did we get it right? The Rolls-Royce/Snecma Olympus 593 Engine reviewed* (https://www.sae.org/technical/papers/912180). SAE International. .

[53] Fisher, S.A.; M.C. Neale and A.J. Brooks. "On the sub-critical stability of variable ramp intakes at Mach numbers around 2" (http://naca.central.cranfield.ac.uk/reports/arc/rm/3711.pdf). *Aeronautical Research Council* (Her Majesty's Stationery Office, 1972). .

[54] "Concorde Special – The test pilot – John Cochrane" (http://www.flightglobal.com/articles/2003/10/21/172657/concorde-special-the-test-pilot-john-cochrane.html). Flight International. 21 October 2003. .

[55] Woodman, Peter (27 July 2000). "How a Concorde pilot would handle a nightmare failure" (http://www.highbeam.com/doc/1G1-63710463.html). Birmingham Post. .

[56] Dalton, Alastair. "Supersonic: The enduring allure of Concorde" (http://www.scotsman.com/news/transport/supersonic-the-enduring-allure-of-concorde-1-2415088). Scotsman.com, 17 July 2012.

[57] Eberhart, Jonathan. "When the SST Is Too Slow..". *Science News* (Society for Science & the Public) **91** (22, 3 June 1967): 528–529. doi:10.2307/3951418. JSTOR 3951418.

[58] "The Concorde takes shape : Test programme and construction proceeding according to schedule" (http://www.emeraldinsight.com/Insight/viewContentItem.do?contentType=Article&hdAction=lnkpdf&contentId=1681441). *Aircraft Engineering and Aerospace Technology* (MCB UP) **38** (4, 1966): 38. 1966. doi:10.1108/eb034143. ISSN 0002-2667. .

[59] N'guyen, V.P.; Perrais; J.P. Perrais (1972). "Fatigue Tests on Big Structure Assemblies of Concorde Aircraft". *Advanced Approaches to Fatigue Evaluation. NASA SP-309* (NASA) **309**: 631. Bibcode 1972NASSP.309..631N.

[60] Wallace, James (7 November 2003). "Those who flew the Concorde will miss it" (http://www.seattlepi.com/business/147276_pilot07.html). Seattle Post. .

[61] Gedge, G.T.; M.I. Prod. "Introduction to Concorde: A brief review of the Concorde and its prospects". *Aircraft Engineering and Aerospace Technology* (Emerald Group Publishing Limited) **40** (3, 1993).

[62] Owen 2001, p. 14.

[63] "Is this the colour of the new millennium?" (http://www.highbeam.com/doc/1P2-4794332.html). UK: *The Independent*, 3 April 1996. .

[64] Frade, Cristina. "Azul contra rojo". *El Mundo*, 5 April 1996.

[65] Owen 2001, p. 78.

[66] Kocivar, Ben. "Aboard the Concorde SST." (http://books.google.co.uk/books?id=lpiMSzja6W4C&lpg=PA142&dq=concorde flexing&pg=PA118#v=onepage&q=concorde flexing&f=false) *Popular Science*, October 1973, p. 117.

[67] "Flight Refuelling Limited and Concorde: The fuel system aboard is largely their work" (http://www.emeraldinsight.com/Insight/viewContentItem.do?contentType=Article&hdAction=lnkpdf&contentId=1682642). *Aircraft Engineering and Aerospace Technology* (MCB UP) **48** (9, September 1976): 20–21. doi:10.1108/eb035344. ISSN 0002-2667. .

[68] "It is the world's most efficient jet engine." -Ricky Bastin, Concorde Engineer

[69] "Concorde model 'B'" (http://www.concordesst.com/concordeb.html)

[70] Strack, William. *Propulsion challenges and opportunities for high-speed transport aircraft* (http://md1.csa.com/partners/viewrecord.php?requester=gs&collection=TRD&recid=N9222540AH&q=Concorde+engines&uid=788858323&setcookie=yes). *Aeropropulsion*, 1987. pp. 437–452. . Retrieved 30 June 2011.

[71] Smale, Alison. "Fuel costs kill Second Generation of Concordes" (http://news.google.co.uk/newspapers?id=Q-0cAAAAIBAJ&sjid=rGcEAAAAIBAJ&pg=6914,3256355&dq=tu+144&hl=en). *Sarasota Herald-Tribune*, 22 September 1979. .

[72] "How much radiation might I be exposed to?" (http://www.britishairways.com/travel/healthcosmic/public/en_gb#4). British Airways. . Retrieved 11 January 2010.

[73] Guerin, D.W. (1973). "Electronic safety test replaces radioactive test source" (http://www.emeraldinsight.com/Insight/viewContentItem.do;jsessionid=E0A02B0587619C9A40D5736FCE7B3F02?contentType=Article&hdAction=lnkpdf&contentId=1682309). *Aircraft Engineering and Aerospace Technology* (*MCB* UP) **45** (4, 1973): 10. doi:10.1108/eb035011. ISSN 0002-2667. .

[74] "Skin cancer danger linked to stratospheric jet planes" (http://news.google.co.uk/newspapers?id=GfUNAAAAIBAJ&sjid=dnkDAAAAIBAJ&pg=2948,4208&dq=second+generation+concorde&hl=en). St. Petersburg Times. 1 April 1975. .

[75] "Cosmic radiation" (http://www.britishairways.com/travel/healthcosmic/public/en_gb). British Airways. . Retrieved 11 January 2010.

[76] Arctowski, Henryk (1940). "On Solar Faculae and Solar Constant Variations" (http://www.pnas.org/cgi/reprint/26/6/406.pdf) (PDF). *Proceedings of the National Academy of Sciences* (*Proc. Natl. Acad. Sci. U.S.A.*, 1940) **26** (6): 406–411. Bibcode 1940PNAS...26..406A. doi:10.1073/pnas.26.6.406. PMC 1078196. PMID 16588370. .

[77] Hepburn, A.N. (1967). "Human Factors in the Concorde" (http://occmed.oxfordjournals.org/cgi/reprint/17/2/47.pdf). *Occupational Medicine* **17** (2, 1967): 47–51. doi:10.1093/occmed/17.2.47. .

[78] Wolff, Mark. "Cabin Decompression and Hypoxia" (http://www.theairlinepilots.com/medical/decompressionandhypoxia.htm). *PIA Air Safety Publication,*6 January 2006. .

[79] *Flight Training Handbook* (http://books.google.com/books?id=ioRTAAAAMAAJ). U.S. Dept. of Transportation, Federal Aviation Administration, Flight Standards Service, 1980. 1980. p. 250. . Retrieved 28 July 2007.

[80] Nunn 1993, p. 341.

[81] Happenny, Steve. "Interim Policy on High Altitude Cabin Decompression – Relevant Past Practice" (http://rgl.faa.gov/Regulatory_and_Guidance_Library\rgPolicy.nsf/0/90AA20C2F35901D98625713F0056B1B8?OpenDocument). *Federal Aviation Administration,* 24 March 2006. .

[82] Schrader 1989, p. 64.

[83] Orlebar 2002, p. 84.

[84] Prestwick Oceanic Area Control Centre: Manual of Air Traffic Services (Part 2). NATS

[85] Orlebar 2002, p. 92.

[86] Orlebar 2002, p. 44.

[87] Schrader 1989, p. 84.

[88] Orlebar 2002, p. 110.

[89] Rose, David. "The real story of Flight 4590: Special Investigation" (http://www.iasa.com.au/folders/Safety_Issues/others/concordespacer.html). *iasa.com.au,* 13 May 2001. .

[90] Brooklands Museum

[91] Stimson, I. L.; Fisher; R. Fisher. "Design and Engineering of Carbon Brakes". *Philosophical Transactions of the Royal Society of London* (The Royal Society) **294** (1411, January 1980): 583–590. Bibcode 1980RSPTA.294..583S. doi:10.1098/rsta.1980.0068. JSTOR 36383.

[92] Owen 2001, p. 118.

[93] "Droop nose" (http://www.flightglobal.com/pdfarchive/view/1971/1971 - 1503.html). Flight International. 12 August 1971. pp. 257–258. .

[94] Owen 2001, p. 84.

[95] "Triplex in Concorde: The story behind the film." (http://www.flightglobal.com/pdfarchive/view/1968/1968 - 2105.html) *Flightglobal.com,* 1968. Retrieved 7 June 2011.

[96] Strang, Dr. W.J; R. McKinley. "Concorde in Service" (http://www.emeraldinsight.com/Insight/viewContentItem.do;jsessionid=2594967949E5659C545A433375AEF406?contentType=Article&hdAction=lnkpdf&contentId=1682798). *Aircraft Engineering and Aerospace Technology* (MCB UP) **50** (12, 1978). doi:10.1108/eb035500. ISSN 0002-2667. .

[97] "BA Tribute to Concorde. The takeoff scene at the end of the video contains a clip of the ATC communication with the "Speedbird Concorde"" (http://video.google.com/videoplay?docid=7422409257931235094). British Airways. . Retrieved 11 January 2010.

[98] Donin, Robert B. *Safety Regulation of the Concorde Supersonic Transport: Realistic Confinement of the National Environmental Policy Act* (http://heinonline.org/HOL/LandingPage?collection=journals&handle=hein.journals/tportl8&div=7&id=&page=). *HeinOnline,* 1976. . Retrieved 30 June 2011.

[99] O'Grady, Jim (27 April 2003). "Neighborhood Report: The Rockaways; Ears Ringing? It's Cheering Over the Demise Of the Concorde" (http://query.nytimes.com/gst/fullpage.html?res=9906E2DC153DF934A15757C0A9659C8B63). *The New York Times,* 27 April 2003. . Retrieved 30 June 2011.

[100] "The Nation: Smooth Landing for the Birds" (http://www.time.com/time/magazine/article/0,9171,915764,00.html). *Time,* 5 December 1977. 5 December 1977. . Retrieved 30 June 2011.

[101] "Concorde facts and figures" (http://www.britishairways.com/concorde/aboutconcorde.html#facts_figures). *British Airways*. . Retrieved 11 January 2010.

[102] Warneke, Ross. "Concorde by June: Offer to Qantas" (http://news.google.co.uk/newspapers?id=3PkQAAAAIBAJ&sjid=hpIDAAAAIBAJ&pg=4864,7521914&dq=concorde+singapore+airlines&hl=en). Australia: *The Age,* 25 October 1977. .

[103] "Singapore Concorde flights" (http://select.nytimes.com/gst/abstract.html?res=F70611FE3C5E167493C6A8178BD95F438785F9). *The New York Times,* 14 October 1977. 14 October 1977. . Retrieved 30 June 2011.

[104] "London and Singapore halt Concorde service" (http://select.nytimes.com/gst/abstract.html?res=F10F15F6395416768FDDAE0994DA415B878BF1D3). *The New York Times,* 17 December 1977. 17 December 1977. . Retrieved 30 June 2011.

[105] "Concorde route cut" (http://news.google.co.uk/newspapers?id=eZckAAAAIBAJ&sjid=raQFAAAAIBAJ&pg=2330,3152191&dq=concorde+singapore+airlines&hl=en). *Montreal Gazette,* 16 September 1980. . Retrieved 30 June 2011.

[106] "French Concorde to Mexico City" (http://news.google.co.uk/newspapers?id=uekpAAAAIBAJ&sjid=4tIEAAAAIBAJ&pg=4845,5120670&dq=concorde+mexico&hl=en). *Daytona Beach Morning Journal,* 11 August 1978. . Retrieved 30 June 2011.

[107] "Supersonic Jet flights suspended" (http://news.google.co.uk/newspapers?id=TuwwAAAAIBAJ&sjid=8t8FAAAAIBAJ&pg=3054,6003139&dq=concorde+mexico&hl=en). *Daytona Beach Morning Journal,* 27 September 1982. . Retrieved 30 June 2011.

[108] "Air France offering 'New Year's Eve in Paris.'" (http://www.highbeam.com/doc/1G1-5212185.html). *PR Newswire,* 2 October 1987. . Retrieved 30 June 2011.

[109] Getze, John (10 February 1977). "Braniff seeks deal to fly Concorde in U.S" (http://pqasb.pqarchiver.com/latimes/access/646052352.html?dids=646052352:646052352&FMT=ABS&FMTS=ABS:AI&type=historic&date=Feb+10,+1977&author=&pub=Los+Angeles+Times&desc=Braniff+Seeks+Deal+to+Fly+Concorde+in+U.S.&pqatl=google). *Los Angeles Times,* 10 February 1977. .

[110] "Concorde flights to Texas Ok'd" (http://pqasb.pqarchiver.com/latimes/access/647634162.html?dids=647634162:647634162&FMT=CITE&FMTS=CITE:AI&type=historic&date=Jun+22,+1978&author=&pub=Los+Angeles+Times&desc=Concorde+Flights+to+Texas+OKd&pqatl=google). *Los Angeles Times,* 22 June 1978. 22 June 1978. . Retrieved 30 June 2011.

[111] "Concorde now reaping profits on N.Y. route" (http://news.google.co.uk/newspapers?id=VAQkAAAAIBAJ&sjid=MO4DAAAAIBAJ&pg=6783,4192544&dq=concorde+braniff&hl=en). *The Spokesman-Review,* 23 November 1979. . Retrieved 30 June 2011.

[112] "Braniff to halt US Concorde flights" (http://news.google.co.uk/newspapers?id=-2UaAAAAIBAJ&sjid=ACsEAAAAIBAJ&pg=6806,4059124&dq=concorde+braniff&hl=en). *Milwaukee Journal,* 16 April 1980. . Retrieved 30 June 2011.

[113] "Concorde flights between Texas and Europe end; Big Dreams at the start, $1,447 for flight to Paris" (http://select.nytimes.com/gst/abstract.html?res=F00911FA3C5C17728DDDA80894DE405B8084F1D3). *The New York Times,* 1 June 1980. 1 June 1980. . Retrieved 30 June 2011.

[114] *B.CAL appoints Concorde consultant.* (http://www.flightglobal.com/pdfarchive/view/1979/1979 - 0919.html) Flight International, 24 March 1979, p. 881.

[115] *Caledonian reports best results.* (http://www.flightglobal.com/pdfarchive/view/1979/1979 - 1617.html) Flight International, 12 May 1979, p. 1547.

[116] *New job for B.CAL Concorde man.* (http://www.flightglobal.com/pdfarchive/view/1979/1979 - 2664.html) Flight International, 21 July 1979, p. 158.

[117] *B.CAL to lease unsold Concorde?.* (http://www.flightglobal.com/pdfarchive/view/1979/1979 - 0541.html) Flight International, 24 February 1979, p. 517.

[118] *Tories support B.CAL's Concorde plan.* (http://www.flightglobal.com/pdfarchive/view/1979/1979 - 1182.html) Flight International, 14 April 1979, p. 1132.

[119] World News. (http://www.flightglobal.com/pdfarchive/view/1979/1979 - 1348.html) Flight International, 28 April 1979, p. 1286.

[120] *High Risk: The Politics of the Air,* Thomson, A., Sidgwick and Jackson, London, 1990, pp. 316–317.

[121] *B.CAL drops Concorde plans but asks for Hong Kong licence.* (http://www.flightglobal.com/pdfarchive/view/1979/1979 - 2433.html) Flight International, 30 June 1979, p. 2331.

[122] *BCAL's African Profit ....* (http://www.flightglobal.com/pdfarchive/view/1971/1971 - 2766.html) World News, Flight International, 23 December 1971, p. 994.

[123] *The Caledonian punchbag,* Flight International, 21 March 1987, p. 33.

[124] *Cheaper fares on Swedish routes?,* Air Transport, Flight International, 31 March 1979, p. 965 (http://www.flightglobal.com/pdfarchive/view/1979/1979 - 1007.html)

[125] *Two Concordes, A310s and 20 new routes in B.CAL package,* Air Transport, Flight International, 19 May 1979, p. 1637 (http://www.flightglobal.com/pdfarchive/view/1979/1979 - 1711.html)

[126] Backroom boys — Francis Spufford

[127] Greenberg, Peter (1 April 1984). "The plane fact is, Concorde has broken the profit barrier for the first time" (http://pqasb.pqarchiver.com/chicagotribune/access/632500842.html?dids=632500842:632500842&FMT=ABS&FMTS=ABS:AI&type=historic&date=Apr+01,+1984&author=&pub=Chicago+Tribune&desc=The+plane+fact+is,+Concorde+has+broken+the+profit+barrier+for+the+first+time&pqatl=google). *Chicago Tribune,* 1 April 1984. . Retrieved 30 June 2011.

[128] "The Concorde belies those who foresaw its extinction" (http://nl.newsbank.com/nl-search/we/Archives?p_product=PI&s_site=philly&p_multi=PI&p_theme=realcities&p_action=search&p_maxdocs=200&p_topdoc=1&p_text_direct-0=0EB29A63D671E93E&p_field_direct-0=document_id&p_perpage=10&p_sort=YMD_date:D&s_trackval=GooglePM). *Philadelphia Inquirer,* 26 January 1986. 26 January 1986. . Retrieved 30 June 2011.

[129] Arnold, James (10 October 2003). "Why economists don't fly Concorde" (http://news.bbc.co.uk/1/hi/business/2935337.stm). *BBC News,* 10 October 2003. .

[130] "Concorde to fly Miami-London route" (http://nl.newsbank.com/nl-search/we/Archives?p_product=MH&s_site=miami&p_multi=MH&p_theme=realcities&p_action=search&p_maxdocs=200&p_topdoc=1&p_text_direct-0=0EB35E9473D0D942&p_field_direct-0=document_id&p_perpage=10&p_sort=YMD_date:D&s_trackval=GooglePM). *Miami Herald,* 24 January 1984. 24 January 1984. . Retrieved 30 June 2011.

[131] "Cuts ground Concorde from Miami to London" (http://nl.newsbank.com/nl-search/we/Archives?p_product=MH&s_site=miami&p_multi=MH&p_theme=realcities&p_action=search&p_maxdocs=200&p_topdoc=1&p_text_direct-0=0EB342A3394157C7&p_field_direct-0=document_id&p_perpage=10&p_sort=YMD_date:D&s_trackval=GooglePM). *Miami Herald.* 13 January 1991. . Retrieved 30 June 2011.

[132] Jensen, Gregory. "After eight years, the Concorde flies to supersonic profit" (http://news.google.com/newspapers?id=R_oxAAAAIBAJ&sjid=NuQFAAAAIBAJ&pg=4247,5250844&dq=concorde+barbados&hl=en). *Reading Eagle,* 27 March 1984. .

[133] Greenberg, Peter S. (23 June 1985). "Flying via charter off to New Bustle" (http://pqasb.pqarchiver.com/latimes/access/64470984.html?dids=64470984:64470984&FMT=ABS&FMTS=ABS:FT&type=current&date=Jun+23,+1985&author=PETER+S.+GREENBERG&pub=Los+Angeles+Times+(pre-1997+Fulltext)&desc=Flying+Via+Charter+Off+to+New+Bustle&pqatl=google). *Los Angeles Times,* 23 June 1985. .

[134] Clark, Jay (18 January 1986). "Supersonic Concorde 10 years in service and still confounding critics" (http://pqasb.pqarchiver.com/thestar/access/474815901.html?dids=474815901:474815901&FMT=ABS&FMTS=ABS:FT&type=current&date=Jan+18,+1986&author=Jay+Clarke+Special+to+The+Star&pub=Toronto+Star&desc=Supersonic+Concorde+10+years+in+service+and+still+confounding+critics&pqatl=google). *Toronto Star,* 18 January 1986. .

[135] "Concorde money is in charters" (http://www.flightglobal.com/pdfarchive/view/1986/1986 - 0201.html). *Flight International,* 25 January 1986. . Retrieved 30 June 2011.

[136] Endres 2001, pp. 110–113.

[137] Cody, E. "French court holds Continental Airlines responsible for 2000 Concorde crash" (http://www.washingtonpost.com/wp-dyn/content/article/2010/12/06/AR2010120601740.html). *Washington Post,* 6 December 2010.

[138] Associated Press (30 November 2012). "French court overturns Concorde crash conviction" (http://www.nzherald.co.nz/world/news/article.cfm?c_id=2&objectid=10850947). New Zealand Herald. . Retrieved 30 November 2012.

[139] "Human Factor Issues Emerge from Concorde Crash Investigation" (http://findarticles.com/p/articles/mi_m0UBT/is_6_16/ai_82753359/). *Air Safety Week, 2002.* . Retrieved 1 February 2010.

[140] "Concorde's safety modifications" (http://news.bbc.co.uk/1/hi/uk/1442061.stm). *BBC News,* 17 July 2001. 17 July 2001. .

[141] "Concorde Completes Successful Test Flight" (http://www.foxnews.com/story/0,2933,29732,00.html). *Fox News,* 17 July 2001. 17 July 2001. . Retrieved 30 June 2011.

[142] "Concorde, 100 BA staff fly over Atlantic" (http://www.highbeam.com/doc/1G1-78118409.html). *United Press International,* 11 September 2001. .

[143] Williams, Timothy. "Concorde returns" (http://news.google.co.uk/newspapers?id=efkjAAAAIBAJ&sjid=yggEAAAAIBAJ&pg=3487,4067442&dq=concorde&hl=en). *Ocala Star-Banner,* 7 November 2001. .

[144] "Concorde 'back where she belongs'" (http://news.bbc.co.uk/1/hi/uk/1641466.stm). *BBC News,* 6 November 2001. 6 November 2001. .

[145] "Concorde grounded for good" (http://news.bbc.co.uk/2/hi/uk_news/2934257.stm). *BBC News,* 10 April 2003. 10 April 2003. . Retrieved 30 June 2011.

[146] Macdonald, Ian S. (1980). "New Aircraft: Where are we heading in the 1980s and 1990s" (http://www.emeraldinsight.com/Insight/viewContentItem.do;jsessionid=E7C9404F2C91DC6FD5BC8C45D43CED69?contentType=Article&hdAction=lnkpdf&contentId=1682944). *Aircraft Engineering and Aerospace Technology* (*MCB UP,*) **52** (7, 1980): 23–17. doi:10.1108/eb035646. ISSN 0002-2667. .

[147] Michaels, Danial (2 October 2003). "Final Boarding Call: As Concorde Departs, so do 3-Man Crews: In New Cockpits, Engineers are seen as Extra Baggage" (http://online.wsj.com/article/SB106504293992222300.html?mod=googlewsj). *Wall Street Journal,* 2 October 2003. .

[148] Montague, Simon (6 May 2003). "Branson's Concorde bid rejected" (http://news.bbc.co.uk/1/hi/uk_politics/3005705.stm). *BBC News,* 6 May 2003. .

[149] "Branson accuses 'sad' Government of washing its hands of Concorde" (http://www.highbeam.com/doc/1G1-109201077.html). *Western Mail,* 24 October 2003. . Retrieved 30 June 2011.

[150] "Concorde not to fly at air shows" (http://edition.cnn.com/2003/WORLD/europe/10/30/concorde.nofly.reut/). *CNN,* 30 October 2003. 30 October 2003. . Retrieved 30 June 2011.

[151] Simpkins, Edward (15 June 2003). "Buffett vehicle to follow in Concorde's slipstream" (http://www.telegraph.co.uk/finance/2855044/Buffett-vehicle-to-follow-in-Concordes-slipstream.html). London: *The Daily Telegraph,* 15 June 2003. .

[152] Quote: "Airbus, the manufacturer of Concorde, has said it is becoming uneconomic to maintain the ageing craft and that it will no longer provide spare parts for it."

[153] "Concorde: An Untimely and Unnecessary Demise" (http://www.thetravelinsider.info/2003/0411.htm). *Travel Insider,* 2003. . Retrieved 13 January 2010.

[154] Lewis, Rob. *Supersonic Secrets: The Unofficial Biography of the Concorde.* London: Expose, a division of Secret Books Limited, 2003. ISBN 0-9546617-0-2.

[155] Pevsner, Donald L.. "The Betrayal of Concorde" (http://www.concorde-spirit-tours.com/concorde.htm). *Concorde Spirit Tours.* . Retrieved 13 January 2010.

[156] "I beat 20m to land last pounds 10 seats on Concorde; Joe grabs bargain of a lifetime." (http://www.thefreelibrary.com/I+BEAT+20m+TO+LAND+LAST+pounds+10+SEATS+ON+CONCORDE;+Joe+grabs...-a061046466) *Scottish Daily Record & Sunday (Glasgow,*

Scotland), 13 February 1997. Retrieved: 13 November 2012.

[157] Lemel, Laurent. "Concorde makes Final Flight from Paris to New York". *Associated Press*, 30 May 2003.

[158] "French Concorde bids adieu" (http://news.bbc.co.uk/1/hi/world/europe/2952594.stm). *BBC News*, 31 May 2003. 31 May 2003. .

[159] "Jetting off" (http://www.highbeam.com/doc/1G1-104484558.html). *Daily Mirror*, 28 June 2003. . Retrieved 30 June 2011.

[160] "3,500 due at UK Concorde auction" (http://news.bbc.co.uk/1/hi/uk/3246326.stm). *BBC News*, 30 November 2003. 30 November 2003. .

[161] "Miscellaneous brief articles – Business & Industry" (http://www.alacrastore.com/storecontent/Business-and-Industry/105669362). *Flight International.* 15 July 2003. . Retrieved 30 June 2011.

[162] "Pourquoi n'a-t-on pas sauvé le Concorde?" (http://www.tourmag.com/Pourquoi-n-a-t-on-pas-sauve-le-Concorde_a7.html). *TourMag.com*, 24 March 2010. . Retrieved 30 June 2011.

[163] Webster, Ben. "This is not a flight of fancy: Volunteers say Concorde can realise an Olympic dream if BA will help". *The Times*, 31 May 2006.

[164] "Museum Sinsheim" (http://sinsheim.technik-museum.de/concorde-f-bvfb/). *Museum Sinsheim.* . Retrieved 26 June 2010.

[165] Atchison, Marc (2 October 2003). "Concorde's supersonic swan song; Star writer aboard for jet's farewell trip to Toronto Transatlantic sound-breaker a vision of grace" (http://pqasb.pqarchiver.com/thestar/access/417072421.html?dids=417072421:417072421&FMT=ABS&FMTS=ABS:FT&type=current&date=Oct+02,+2003&author=Marc+Atchison&pub=Toronto+Star&desc=Concorde's+supersonic+swan+song;+Star+writer+aboard+for+jet's+farewell+trip+to+Toronto+Transatlantic+sound-breaker+a+vision+of+grace&pqatl=google). *Toronto Star*, 2 October 2003. .

[166] Ramstack, Tom. "Final flight: British Airways Concorde lands locally for last time" (http://goliath.ecnext.com/coms2/gi_0199-3250817/Final-flight-British-Airways-Concorde.html). *Washington Post*, 15 October 2003. .

[167] "Concorde establishes London-to-U.S. record" (http://www.accessmylibrary.com/coms2/summary_0286-24654137_ITM). *America's Intelligence Wire*, 9 October 2003. 9 October 2003. . Retrieved 30 June 2011.

[168] "SST makes record flight" (http://nl.newsbank.com/nl-search/we/Archives?p_product=SL&p_theme=sl&p_action=search&p_maxdocs=200&p_topdoc=1&p_text_direct-0=0EB04F816AA79462&p_field_direct-0=document_id&p_perpage=10&p_sort=YMD_date:D&s_trackval=GooglePM). *St Louis Post*, 9 February 1996. 9 February 1996. . Retrieved 30 June 2011.

[169] "Exhibits: Concorde". (http://www.intrepidmuseum.org/The-Intrepid-Experience/Exhibits/Concorde.aspx) *Intrepidmuseum.org.* Retrieved 6 March 2011.

[170] "Last Concorde lands" (http://news.bbc.co.uk/2/hi/uk_news/england/bristol/somerset/3238674.stm). *BBC News*,. 27 November 2003. . Retrieved 30 June 2011.

[171] "Concorde enjoys Cardiff farewell" (http://news.bbc.co.uk/1/hi/wales/3205915.stm). *BBC News.* 23 October 2003. 23 October 2003. .

[172] "Concorde – The Farewell – A collection of the final flights of the last days of Concorde" (http://www.amazon.co.uk/dp/B0001DI57G). *Simply Media*, 19 April 2004. . Retrieved 30 June 2011.

[173] Magoolaghan, Brian. "The Concorde Makes A Comeback" (http://www.rockawave.com/news/2003-10-31/Front_Page/028.html). *Wave of Long Island*, 31 October 2003. . Retrieved 30 June 2011.

[174] Lawless, Jill. "Final Concorde flight lands at Heathrow". *Associated Press*, 26 October 2003.

[175] "End of an era for Concorde" (http://news.bbc.co.uk/1/hi/uk/3211053.stm). *BBC News*,. 24 October 2003. .

[176] Thousands welcome Concorde to Seattle (http://www.seattlepi.com/local/article/Thousands-welcome-Concorde-to-Seattle-1128999.php)

[177] Robert Bogash (http://rbogash.com/Concorde.html)

[178] "BBC News: Last Concorde lands" (http://news.bbc.co.uk/1/hi/england/bristol/3238674.stm). *BBC News*,. 27 November 2003. .

[179] "Will Concorde ever come out of retirement – e.g. for a Coronation flypast or airshows?" (http://www.britishairways.com/concorde/faq.html#17). *British Airways*. . Retrieved 14 January 2010.

[180] "Concorde nose cone sells for half-million at auction" (http://www.usatoday.com/travel/news/2003-12-02-concorde_x.htm). *USA Today*, 12 December 2003. 2 December 2003. . Retrieved 30 June 2011.

[181] "Concorde Memorabilia Auction" (http://www.highbeam.com/doc/1P1-87889450.html). *Getty Images*, 1 December 2003. . Retrieved 30 June 2011.

[182] "Heathrow Concorde model removed" (http://news.bbc.co.uk/1/hi/england/london/6509667.stm). *BBC News*,. 30 March 2007. .

[183] "The Brooklands Concorde Project" (http://www.concordeproject.com/). *Brooklands Museum*. . Retrieved 15 January 2010.

[184] "Air France Concorde to taxi again under own power" (http://www.flightglobal.com/articles/2010/02/05/338114/air-france-concorde-to-taxi-again-under-own-power.html). *Flightglobal*, 5 February 2010. . Retrieved 5 February 2010.

[185] "Work starts in £15m plan to get Concorde Flying" (http://news.bbc.co.uk/1/hi/world/europe/8712806.stm). *BBC News*,29 May 2010. 29 May 2010. . Retrieved 29 May 2010.

[186] "Soviet Union: Christening the Concordski" (http://www.time.com/time/magazine/article/0,9171,912014,00.html). *Time*, 14 November 1977. 14 November 1977. . Retrieved 30 June 2011.

[187] Gordon, Yefim. *Tupolev Tu-144.* London: Midland, 2006. ISBN 1-85780-216-0.

[188] "'Concordski' designer dies" (http://news.bbc.co.uk/1/hi/world/europe/1328132.stm). *BBC News*, 13 May 2001. 13 May 2001. .

[189] Hess, John L (26 May 1971). "Soviet SST, in Its First Flight to the West, Arrives in Paris for Air Show" (http://select.nytimes.com/gst/abstract.html?res=F4091EFB3954127B93C4AB178ED85F458785F9). *The New York Times*, 26 May 1971. . Retrieved 30 June 2011.

[190] Deruaz, George. "Soviet SST stalls, dives into towns" (http://news.google.co.uk/newspapers?id=2JwMAAAAIBAJ&sjid=5l8DAAAAIBAJ&pg=4974,3987586&dq=tu+144&hl=en). *St. Petersburg Times,* 4 June 1973. . Retrieved 30 June 2011.
[191] "Pride of Soviet air fleet explodes during exhibition" (http://news.google.co.uk/newspapers?id=DBsfAAAAIBAJ&sjid=Ko0EAAAAIBAJ&pg=3185,500285&dq=tu+144&hl=en). *Sarasota Journal,* 4 June 1973. . Retrieved 30 June 2011.
[192] Fisher, Dan (27 October 1978). "Russia confirms crash of Supersonic Airliner in test: Latest failure of Trouble-plagued TU-144 seen as blow to Soviet hopes of expanding industry" (http://pqasb.pqarchiver.com/latimes/access/659730132.html?dids=659730132:659730132&FMT=ABS&FMTS=ABS:AI&type=historic&date=Oct+27,+1978&author=&pub=Los+Angeles+Times&desc=Russia+Confirms+Crash+of+Supersonic+Airliner+in+Test&pqatl=google). *Los Angeles Times,* 27 October 1978. .
[193] "ASN Aircraft accident Tupolev 144D CCCP-77111 Yegoryevsk" (http://aviation-safety.net/database/record.php?id=19780523-1). Flight Safety Foundation. . Retrieved 1 July 2011.
[194] "The Tu-144: the future that never was" (http://en.rian.ru/analysis/20080603/109170103.html). *RIA Novosti,* 3 January 2008. . Retrieved 30 June 2011.
[195] Fridlyander, Iosif. "Sad Epic of the Tu-144." *Messenger of Russian Academy of Sciences*, №1, 2002 (in Russian: И.Н. Фридляндер, "Печальная эпопея Ту-144" (http://vivovoco.rsl.ru/VV/JOURNAL/VRAN/02_01/FRID.HTM), Вестник РАН, №1, 2002.
[196] "The United States SST Contenders" (http://www.flightglobal.com/pdfarchive/view/1964/1964 - 0377.html). *Flight International,* 13 February 1964. pp. 234–235. . Retrieved 30 June 2011.
[197] Winchester 2005, p. 84.
[198] Lyons, Richard D (5 January 1969). "The Russians Lead With the SST..." (http://select.nytimes.com/gst/abstract.html?res=F10D11F6395F127A93C7A9178AD85F4D8685F9). *The New York Times,* 5 January 1969. . Retrieved 30 June 2011.
[199] "B-58's Sonic Boom Rattles Kentuckians" (http://pqasb.pqarchiver.com/chicagotribune/access/574189232.html?dids=574189232:574189232&FMT=CITE&FMTS=CITE:AI&date=Dec+19,+1961&author=&pub=Chicago+Daily+Tribune&desc=B-58's+SONIC+BOOM+RATTLES+KENTUCKIANS&pqatl=google). *Chicago Daily Tribune,* 19 December 1961. 19 December 1961. . Retrieved 30 June 2011.
[200] "The Nation: Showdown on the SST" (http://www.time.com/time/magazine/article/0,9171,944291,00.html). *Time,* 29 May 1971. 29 March 1971. . Retrieved 30 June 2011.
[201] "Tu-160 Blackjack Strategic Bomber, Russia" (http://www.airforce-technology.com/projects/tu160/). *airforce-technology.com.* . Retrieved 16 May 2010.
[202] "Here Comes the Concorde, Maybe" (http://www.time.com/time/magazine/article/0,9171,918012,00.html). *Time,* 16 February 1976. 16 February 1976. . Retrieved 30 June 2011.
[203] Allen, Robert M.. *Legal and Environmental ramifications of the Concorde* (http://heinonline.org/HOL/LandingPage?collection=journals&handle=hein.journals/jalc42&div=28&id=&page=). J. Air L. & Com., 1976. . Retrieved 30 June 2011.
[204] Hock, R.; R. Hawkins. *Recent studies into Concorde noise reduction* (http://md1.csa.com/partners/viewrecord.php?requester=gs&collection=TRD&recid=N7422661AH&q=Concorde+noise&uid=788872723&setcookie=yes). *AGARD Noise Mech,* 1974. p. 14. . Retrieved 30 June 2011.
[205] Muss, Joshua A.. *Aircraft Noise: Federal pre-emption of Local Control, Concorde and other recent cases* (http://heinonline.org/HOL/LandingPage?collection=journals&handle=hein.journals/jalc43&div=45&id=&page=). *J. Air L. & Com,* 1977. . Retrieved 30 June 2011.
[206] Endres 2001, p 90.
[207] "Reducing noise with type 28 nozzle" (http://www.emeraldinsight.com/Insight/viewContentItem.do?contentType=Article&hdAction=lnkpdf&contentId=1682311). *Aircraft Engineering and Aerospace Technology* (*MCB UP*) **45** (4, 1973): 14. 1973. doi:10.1108/eb035013. .
[208] Fahey, D. W.; et al. (1995). "Emission Measurements of the Concorde Supersonic Aircraft in the Lower Stratosphere" (http://www.sciencemag.org/cgi/content/abstract/sci;270/5233/70). *Science* **270** (5233): 70. Bibcode 1995Sci...270...70F. doi:10.1126/science.270.5233.70. .
[209] "Increase in supersonic jets could be threat to ozone U-2 plane trails Concorde, studies exhaust particles." (http://articles.baltimoresun.com/1995-10-08/news/1995281022_1_ozone-sulfur-exhaust-particles) *Baltimore Sun,* October 8, 1995.
[210] Anderson, Jon. *Decision Analysis in Environmental Decisionmaking: Improving the Concorde Balance* (http://heinonline.org/HOL/LandingPage?collection=journals&handle=hein.journals/cjel5&div=9&id=&page=). *HeinOnline,* 1978. . Retrieved 30 June 2011.
[211] "Train à grande vitesse causes distress" (http://www.esemag.com/1101/1101ed.html). *Environmental Science and Engineering,* November 2001. .
[212] "National and regional tranquillity maps" (http://www.cpre.org.uk/campaigns/landscape/tranquillity/national-and-regional-tranquillity-maps). *Campaign to Protect Rural England.* . Retrieved 25 April 2010.
[213] "British Airways Concorde is expected to begin flying passengers again in next 6 weeks" (http://www.alacrastore.com/storecontent/Business-and-Industry/24962872). *Dallas Morning News,* 23 August 2001. . Retrieved 30 June 2011.
[214] "Farewell to Concorde" (http://news.bbc.co.uk/1/hi/in_depth/uk/2003/concorde_retirement/default.stm). *BBC News,* 15 August 2007. 15 August 2007. . Retrieved 15 January 2010.
[215] Oxford Language Dictionaries Online – French Resources (http://www.oxfordlanguagedictionaries.com/Public/PublicResources.html?direction=b-fr-en&sp=S/oldo/resources/fr/French-supp-05.xml): Glossary of Grammatical Terms
[216] Centre National de Ressources Textuelles et Lexicales (http://www.cnrtl.fr/lexicographie/academie9/le) – Définition de LE, LA: article défini, II.3

[217] Reverso Dictionnaire: La majuscule dans les noms propres (http://grammaire.reverso.net/5_2_01_la_majuscule_dans_les_noms_propres.shtml) ("Capital letters within proper names")

[218] **concorde** *s.f.* concord, unity, harmony, peace.Ferrar 1980, p. 114.

[219] "Celebrate Concorde: videos" (http://www.britishairways.com/travel/concvidhome/public/en_gb). *British Airways*. . Retrieved 15 January 2010.

[220] "Documentary of British Airways Concorde introduction" (http://www.youtube.com/watch?v=Ro3QVMCG-aY). *youtube.com*. .

[221] Quote: Raymond Baxter commentating as Concorde flies for first time: "She rolls ... She flies!"

[222] "Red Arrows fly into Scotland" (http://www.highbeam.com/doc/1G1-62676656.html). *Daily Record,* 12 June 2000. . Retrieved 30 June 2011.

[223] "Million turn out to crown Queen's Jubilee" (http://www.telegraph.co.uk/education/3297816/Million-turn-out-to-crown-Queens-Jubilee.html). London: *Daily Telegraph,* 4 June 2002. 4 June 2002. . Retrieved 1 April 2010.

[224] Laville, Sandra (24 October 2003). "Chaos fear at Concorde farewell" (http://www.telegraph.co.uk/news/uknews/1444939/Chaos-fear-at-Concorde-farewell.html). London: *Daily Telegraph,* 24 October 2003. .

[225] "Air France Concorde sets round-the-world speed record" (http://www.allbusiness.com/transportation/air-transportation-airports/7142582-1.html). *Business Wire,* 16 August 1995. . Retrieved 30 June 2011.

[226] "French Concorde to attempt round-the-world record". *Anchorage Daily News,* 12 October 1992.

[227] "Concorde jets occupants on record ride". *Deseret News,* 17 August 1995.

[228] "Rolls-Royce SNECMA Olympus". *Janes,* 25 July 2000.

[229] "Museum of Flight" (http://www.museumofflight.org/). Museum of Flight. . Retrieved 21 August 2011.

[230] "French transport chief speculates about new-generation Concorde" (http://nl.newsbank.com/nl-search/we/Archives?p_product=NewsLibrary&p_multi=APAB&d_place=APAB&p_theme=newslibrary2&p_action=search&p_maxdocs=200&p_topdoc=1&p_text_direct-0=0F89C550A53D43FD&p_field_direct-0=document_id&p_perpage=10&p_sort=YMD_date:D&s_trackval=GooglePM). *Associated Press,* 17 August 2000. 17 August 2000. . Retrieved 30 June 2011.

[231] Cody, Edward (10 May 1990). "Partnership gears up for Concorde sequel; British, French firms sign plane pact" (http://pqasb.pqarchiver.com/washingtonpost/access/72587838.html?dids=72587838:72587838&FMT=ABS&FMTS=ABS:FT&type=current&date=May+10,+1990&author=Edward+Cody&pub=The+Washington+Post+(pre-1997+Fulltext)&desc=Partnership+Gears+Up+For+Concorde+Sequel;+British,+French+Firms+Sign+Plane+Pact&pqatl=google). *Washington Post,* 10 May 1990. .

[232] "Firm considers 'son of Concorde'" (http://news.bbc.co.uk/1/hi/business/3231354.stm). *BBC News,* 23 November 2003. 23 November 2003. .

[233] "Japan, France working on new supersonic jet" (http://www.msnbc.msn.com/id/8229113/). *MSNBC,* 15 June 2005. . Retrieved 30 June 2011.

[234] "Japan tests supersonic jet model" (http://news.bbc.co.uk/2/hi/science/nature/4325634.stm). *BBC News,* 10 October 2005. 10 October 2005. . Retrieved 30 June 2011.

[235] Kaminski-Morrow, David (19 June 2011). "Paris: EADS details near-hypersonic transport concept" (http://www.flightglobal.com/articles/2011/06/19/358271/paris-eads-details-near-hypersonic-transport-concept.html). Flight Daily News. .

[236] Lichfield, John (20 June 2011). "London to New York in 90 minutes: is this the Concorde of the future?" (http://www.independent.co.uk/news/science/london-to-new-york-in-90-minutes-is-this-the-concorde-of-the-future-2299925.html). *The Independent*. .

[237] "LAPCAT aims at supersonic civil aviation" (http://www.gizmowatch.com/entry/lapcat-aims-at-supersonic-civil-aviation/). *Gizmo Watch,* 30 August 2007. . Retrieved 3 July 2009.

[238] "LAPCAT II – Facts and figures" (http://www.esa.int/esaMI/Space_Engineering/SEMD0G5KXMF_0.html). *European Space Agency,* 17 November 2009. . Retrieved 10 August 2010.

[239] O'Connell, Dominic (18 May 2008). "Orders for Aerion's Concorde executive jet are more than $3 billion" (http://business.timesonline.co.uk/tol/business/industry_sectors/transport/article3953823.ece). London: *The Times,* 18 May 2008. .

[240] Nanveen. "More details emerge on the $80 million Aerion Supersonic Business Jet." (http://www.myduckeggs.com/category/supersonic-aerospace-international/) *myduckeggs.com,* 26 July 2010. Retrieved 28 July 2010.

[241] Hagerman, Eric. "Supersonic jet promises to fly nearly silent" (http://www.cnn.com/2007/TECH/02/16/quiet.supersonic/index.html). *CNN,* 16 February 2007. .

[242] Kelly 2005, p. 52.

[243] "Comparison with Tu-144." (http://www.richard-seaman.com/Aircraft/Museums/Monino/Highlights/index.html) *Richard Seaman aircraft museum.* Retrieved 15 January 2010.

[244] BA and AF Concordes originally had 100 seats, with AF removing 8 seats after the safety modifications of CY2000–2001 because of weight considerations.

Citations

Bibliography

- Beniada, Frederic. *Concorde.* Minneapolis, Minnesota: Zenith Press, 2006. ISBN 0-7603-2703-3.
- Calvert, Brian. *Flying Concorde: The Full Story.* London: Crowood Press, 2002. ISBN 1-84037-352-0.
- Deregel, Xavier; Jean-Philippe Lemaire. *Concorde Passion.* New York: LBM, 2009. ISBN 2-915347-73-5.

- Endres, Günter. *Concorde*. St Paul, Minnesota: MBI Publishing Company, 2001. ISBN 0-7603-1195-1.
- Ferrar, Henry (ed.). *The Concise Oxford French-English Dictionary*. New York: Oxford University Press, 1980. ISBN 0-19-864157-5.
- Frawley, Gerald (2003). *The International Directory of Civil Aircraft, 2003/2004*. Aerospace Publications. ISBN 1-875671-58-7.
- Gunn, John (2010). *Crowded Skies*. Turnkey Productions. ISBN 978-0-646-54973-6.
- Kelly, Neil. *The Concorde Story: 34 Years of Supersonic Air Travel*. Surrey, UK: Merchant Book Company Ltd, 2005. ISBN 1-904779-05-0.
- Knight, Geoffrey. *Concorde: The Inside Story*. London: Weidenfeld and Nicolson, 1976. ISBN 0-297-77114-0.
- McIntyre, Ian. *Dogfight: The Transatlantic Battle over Airbus*. Westport, Connecticut: Praeger Publishers, 1992. ISBN 0-275-94278-3.
- Nunn, John Francis. *Nunn's Applied Respiratory Physiology*. Burlington, Maryland: Butterworth-Heineman, 1993. ISBN 0-7506-1336-X.
- Owen, Kenneth (2001). Concorde: Story of a Supersonic Pioneer (http://books.google.com/?id=VzNUJlX7CXoC&printsec=frontcover&dq=Concorde:+story+of+a+supersonic+pioneer#v=onepage&q&f=false). London: Science Museum, 2001. ISBN 978-1-900747-42-4.
- Orlebar, Christopher. *The Concorde Story*. Oxford, UK: Osprey Publishing, 2004. ISBN 1-85532-667-1.
- Ross, Douglas (March 1978). *The Concorde Compromise: The Politics of Decision-making*. Bulletin of the Atomic Scientists. pp. 46–53.
- Schrader, Richard K. *Concorde: The Full Story of the Anglo-French SST*. Kent, UK: Pictorial Histories Pub. Co., 1989. ISBN 0-929521-16-1.
- Taylor, John W. R. (1965). *Jane's All The World's Aircraft 1965–66*. Marston.
- Towey, Barrie (ed.). *Jet Airliners of the World 1949–2007*. Tunbridge Wells, Kent, UK: Air-Britain (Historians) Ltd, 2007. ISBN 0-85130-348-X.
- Winchester, Jim. *The World's Worst Aircraft: From Pioneering Failures to Multimillion Dollar Disasters*. London: Amber Books Ltd, 2005. ISBN 1-904687-34-2.
- Winchester, Jim (2005). *X-Planes and Prototypes: From Nazi Secret Weapons to the Warplanes of the Future*. Amber Books Ltd. ISBN 1-84013-815-7.

External links

- British Airways Concorde page (http://www.britishairways.com/concorde/index.html)
- Braniff Airways Concorde page (http://www.braniffpages.com/concorde.html)
- Design Museum (UK) Concorde page (http://www.designmuseum.org/design/concorde)
- Concorde Experience at the National Museum of Flight, Scotland (http://www.nms.ac.uk/our_museums/museum_of_flight/concorde.aspx)
- Concorde F-BVFB at the Auto & Technik Museum Sinsheim (http://sinsheim.technik-museum.de/en/concorde-f-bvfb)
- Supersonic: Tribute to Concorde (http://www.life.com/image/first/in-gallery/24961/supersonic-tribute-to-concorde) – a slideshow by *Life magazine*
- Pathé film of 002's maiden flight from Filton to RAF Fairford on 9 April 1967 (http://www.britishpathe.com/record.php?id=45713)
- First Supersonic Transatlantic Commercial Flight Lands at Dulles (1976) (http://ghostsofdc.org/2012/01/30/first-supersonic-transatlantic-commercial-flight-lands-at-dulles-1976/)

Concorde

Concorde	

Role	Supersonic airliner
Manufacturer	BAC (now BAE Systems) Sud-Aviation, Aérospatiale (now EADS)
First flight	2 March 1969
Introduction	21 January 1976
Retired	26 November 2003
Status	Retired from service
Primary users	British Airways Air France See *Operators* below for others
Number built	20 (including 6 non-airline aircraft)[1] [2]
Program cost	£1.3 billion[3]
Unit cost	£23 million in 1977

Aérospatiale-BAC Concorde (pron.: /ˈkɒŋkɔrd/) is a retired turbojet-powered supersonic passenger airliner or supersonic transport (SST). It is the second of only two SSTs to have entered commercial service; the first was the Tupolev Tu-144. Concorde was a product of the manufacturing efforts of Aérospatiale and the British Aircraft Corporation (BAC), produced under a joint Franco-British treaty. First flown in 1969, Concorde entered service in 1976 and continued commercial flights for 27 years.

Among other destinations, Concorde flew regular transatlantic flights from London Heathrow and Paris-Charles de Gaulle Airport to New York JFK and Washington Dulles; it profitably flew these routes in less than half the time of other airliners. With only 20 aircraft built, the development of Concorde was a substantial economic loss; Air France and British Airways also received considerable government subsidies to purchase them. Concorde was retired in 2003 due to a general downturn in the aviation industry after the type's only crash in 2000, the 9/11 terrorist attacks in 2001 and a decision by Airbus, the successor firm of Aerospatiale and BAC, to discontinue maintenance support.[4]

Concorde's name reflects the development agreement between the United Kingdom and France. In the UK, any or all of the type—unusual for an aircraft—are known simply as "Concorde", without an article. The aircraft is regarded by many people as an aviation icon and an engineering marvel.[5]

Development

Concept

In the late 1950s, France, the United Kingdom, United States, and Soviet Union were considering developing supersonic transport. The British Bristol Aeroplane Company and the French Sud Aviation were both working on designs, called the Type 223 and Super-Caravelle, respectively. Both were largely funded by their respective governments.[6] The British design was for a thin-winged delta shape[7] (which owed much to work by Dietrich Küchemann, then at the Royal Aircraft Establishment) for a transatlantic-ranged aircraft for about 100 people, while the French were intending to build a medium-range aircraft.[6]

Concorde's final flight; G-BOAF from Heathrow to Bristol, on 26 November 2003. The extremely high fineness ratio of the fuselage is evident.

The designs were both ready to start prototype construction in the early 1960s, but the cost was so great that the British government made it a requirement that British Aircraft Corporation (which had been formed in 1960 as a consolidation of British aircraft companies, including the Bristol Aeroplane Company) look for international co-operation.[6] Approaches were made to a number of countries, but only France showed real interest. The development project was negotiated as an international treaty between the two countries rather than a commercial agreement between companies and included a clause, originally asked for by the UK, imposing heavy penalties for cancellation. A draft treaty was signed on 29 November 1962. By this time, both companies had been merged into new ones; thus, the Concorde project was between the British Aircraft Corporation and Aérospatiale.[6] At first the new consortium intended to produce one long range and one short range version. However, prospective customers showed no interest in the short-range version and it was dropped.[6] The consortium secured orders (i.e., non-binding options) for over 100 of the long-range version from the major airlines of the day: Pan Am, BOAC, and Air France were the launch customers, with six Concordes each. Other airlines in the order book included Panair do Brasil, Continental Airlines, Japan Airlines, Lufthansa, American Airlines, United Airlines, Air India, Air Canada, Braniff, Singapore Airlines, Iran Air, Olympic Airways, Qantas, CAAC, Middle East Airlines, and TWA.[6] [8] [9]

The design work was supported by a preceding research programme studying the flight characteristics of low ratio delta wings. The supersonic BAC 221 was modified for flight tests of the high speed flight envelope,[10] the Handley Page HP.115 also provided valuable information on low speed performance.[11]

Naming

Reflecting the treaty between the French and British governments which led to Concorde's construction, the name *Concorde* is from the French word *concorde* (IPA: [kɔ̃kɔʁd]), which has an English equivalent, *concord*. Both words mean *agreement*, *harmony* or *union*.

British Airways Concorde in early BA livery at London Heathrow Airport, in the 1980s

The aircraft was initially referred to in the UK as *Concorde*, with the French spelling, but was officially changed to *Concord* by Harold Macmillan in response to a perceived slight by Charles de Gaulle.[12] In 1967, at the French roll-out in Toulouse the British Government Minister for Technology, Tony Benn, announced that he would change the spelling back to *Concorde*.[12] This created a nationalist uproar that died down when Benn stated that the suffixed e represented "Excellence, England, Europe and Entente (Cordiale)." In his memoirs, he recounts a tale of a letter from an irate Scotsman claiming: "[Y]ou talk about 'E' for England, but part of it is made in Scotland." Given Scotland's contribution of providing the nose cone for the aircraft, Benn replied, "[I]t was also 'E' for 'Écosse' (the French name for Scotland) — and I might have added 'e' for extravagance and 'e' for escalation as well!"[13]

Concorde also acquired an unusual nomenclature for an aircraft. In common usage in the United Kingdom, the type is known as *Concorde* without an article, rather than ***the*** *Concorde* or ***a*** *Concorde*.[14] [15]

Testing

Construction of two prototypes began in February 1965: 001, built by Aerospatiale at Toulouse, and 002, by BAC at Filton, Bristol. Concorde 001 made its first test flight from Toulouse on 2 March 1969, piloted by André Turcat,[16] and first went supersonic on 1 October.[17] The first UK-built Concorde flew from Filton to RAF Fairford on 9 April 1969, piloted by Brian Trubshaw.[18] [19] Both prototypes were presented to the public for the first time on 7–8 June 1969 at the Paris Airshow. As the flight programme progressed, 001 embarked on a sales and demonstration tour on 4 September 1971, which was also the first transatlantic crossing of Concorde.[20] [21] Concorde 002 followed suit on 2 June 1972 with a tour of the Middle and Far East.[22] Concorde 002 made the first visit to the United States in 1973, landing at the new Dallas/Fort Worth Regional Airport to mark that airport's opening.[23]

Concorde's first visit to Heathrow Airport on 1 July 1972.

While Concorde had initially held a great deal of customer interest, the project was hit by a large number of order cancellations. The Paris Le Bourget air show crash of the competing Soviet Tupolev Tu-144 had shocked potential buyers, and public concern over the environmental issues presented by a supersonic aircraft – the sonic boom, takeoff-noise and pollution – had produced a shift in public opinion of SSTs. By 1976 four nations remained as prospective buyers: Britain, France, China, and Iran.[24] Only Air France and British Airways (the successor to BOAC) took up their orders, with the two governments taking a cut of any profits made.[25]

The United States cancelled the Boeing 2707, its rival supersonic transport programme, in 1971. Observers have suggested that opposition to Concorde on grounds of noise pollution had been encouraged by the United States Government, as it lacked its own competitor.[26] The US, India, and Malaysia all ruled out Concorde supersonic flights over the noise concern, although some of these restrictions were later relaxed.[27] [28] Professor Douglas Ross characterised restrictions placed upon Concorde operations by President Jimmy Carter's administration as having been an act of protectionism of American aircraft manufacturers.[29]

Concorde had other considerable difficulties that led to its dismal sales performance. Costs had spiralled during development to more than six times the original projections, arriving at a unit cost of £23 million in 1977.[30] World events had also dampened Concorde sales prospects, the 1973 oil crisis had many airlines think twice about aircraft with high rates of fuel-consumption; and new wide-body aircraft, such as the Boeing 747, had recently made subsonic aircraft significantly more efficient and presented a low-risk option for airlines.[31] While carrying a full load, Concorde achieved 15.8 passenger miles per gallon of fuel, while the Boeing 707 reached 33.3 pm/g, the Boeing 747 46.4 pm/g, and the McDonnell Douglas DC-10 53.6 pm/g.[32] An emerging trend in the industry in favour of cheaper airline tickets had also caused airlines such as Qantas to question Concorde's market suitability.[33]

Design

General features

Concorde is an ogival (also "ogee") delta-winged aircraft with four Olympus engines based on those employed in the RAF's Avro Vulcan strategic bomber. Concorde was the first airliner to have a (in this case, analogue) fly-by-wire flight-control system; the avionics of Concorde were unique because it was the first commercial aircraft to employ hybrid circuits.[34] The principal designer for the project was Pierre Satre, with Sir Archibald Russell as his deputy.[35]

Concorde cockpit layout

Concorde pioneered the following technologies:

For high speed and optimisation of flight:

- Double delta (ogee/ogival) shaped wings[7]
- Variable engine air intake system controlled by digital computers[36]
- Supercruise capability[37]
- Thrust-by-wire engines, predecessor of today's FADEC-controlled engines[36]
- Droop-nose section for better landing visibility

For weight-saving and enhanced performance:

- Mach 2.04 (~2179 kilometres per hour / mph) cruising speed[38] for optimum fuel consumption (supersonic drag minimum although turbojet engines are more efficient at higher speed[39])
- Mainly aluminium construction for low weight and conventional manufacture (higher speeds would have ruled out aluminium)[40]
- Full-regime autopilot and autothrottle[41] allowing "hands off" control of the aircraft from climb out to landing
- Fully electrically controlled analogue fly-by-wire flight controls systems[34]
- High-pressure hydraulic system of 28 MPa (4,000 lbf/in²) for lighter hydraulic components[42]
- Complex Air Data Computer (ADC) for the automated monitoring and transmission of aerodynamic measurements (total pressure, static pressure, angle of attack, side-slip).[43]
- Fully electrically controlled analogue brake-by-wire system[44]
- Pitch trim by shifting fuel around the fuselage for centre-of-gravity control[45]
- Parts made using "sculpture milling", reducing the part count while saving weight and adding strength.[46]
- No auxiliary power unit, as Concorde would only visit large airports where ground air start carts are available.[47]

Engines

Close up of pre-production engine nozzles of G-AXDN. The nozzle/thrust reverser design was altered for production

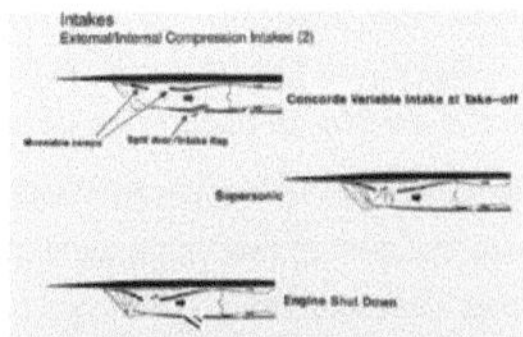

Concorde's intake system schematics

Concorde's intake system

Concorde needed to fly long distances to be economically viable; this required high efficiency. Turbofan engines were rejected due to their larger cross-section producing excessive drag. Turbojets were found to be the best choice of engines.[48] The engine used was the twin spool Rolls-Royce/Snecma Olympus 593, a development of the Bristol engine first used for the Avro Vulcan bomber, and developed into an afterburning supersonic variant for the BAC TSR-2 strike bomber.[49]

The aircraft used reheat (afterburners) at takeoff and to pass through the transonic regime (i.e., "go supersonic") between Mach 0.95 and Mach 1.7. The afterburners were switched off at all other times.[50] Due to jet engines being highly inefficient at low speeds, Concorde burned two tonnes of fuel (almost 2% of the maximum fuel load) taxiing to the runway.[51] To conserve fuel only the two outer engines were run after landing for taxiing.

The intake design for Concorde's engines was especially critical.[52] Conventional jet engines can take in air at only around Mach 0.5; therefore the air has to be slowed from the Mach 2.0 airspeed that enters the engine intake. In particular, Concorde needed to control the shock waves that this reduction in speed generates to avoid damage to the engines. This was done by a pair of intake ramps and an auxiliary spill door, whose position moved in-flight to slow transiting air.[53]

Engine failure causes problems on conventional subsonic aircraft; not only does the aircraft lose thrust on that side but the engine creates drag, causing the aircraft to yaw and bank in the direction of the failed engine. If this had happened to Concorde at supersonic speeds, it theoretically could have caused a catastrophic failure of the airframe. Although computer simulations predicted considerable problems, in practice Concorde could shut down both engines on the same side of the aircraft at Mach 2 without the predicted difficulties.[54] During an engine failure the required air intake is virtually zero so, on Concorde, engine failure was countered by the opening of the auxiliary spill door and the full extension of the ramps, which deflected the air downwards past the engine, gaining lift and minimising drag. Concorde pilots were routinely trained to handle double engine failure.[55]

Heating issues

Air friction on the outer surfaces caused the cabin to heat up during flight. Every surface, such as windows and panels, was warm to the touch by end of the flight.[56] Besides engines, the hottest part of the structure of any supersonic aircraft, due to aerodynamic heating, is the nose. The engineers used Hiduminium R.R. 58, an aluminium alloy, throughout the aircraft due to its familiarity, cost and ease of construction. The highest temperature that aluminium could sustain over the life of the aircraft was 127 °C (°F), which limited the top speed to Mach 2.02.[57] Concorde went through two cycles of heating and cooling during a flight, first cooling down as it gained altitude, then heating up after going supersonic. The reverse happened when descending and slowing down. This had to be factored into the metallurgical and fatigue modelling. A test rig was built that repeatedly heated up a full-size section of the wing, and then cooled it, and periodically samples of metal were taken for testing.[58] [59]

Owing to air friction as the plane traveled at supersonic speed, the fuselage would heat up and expand by as much as 300 mm (almost 1 ft). The most obvious manifestation of this being a gap that opened up on the flight deck between the flight engineer's console and the bulkhead. On some aircraft that conducted a retiring supersonic flight, the flight engineers placed their caps in this expanded gap, wedging the cap when it shrank again.[60] To keep the cabin cool, Concorde used the fuel as a heat sink for the heat from the air conditioning.[61] The same method also cooled the hydraulics. During supersonic flight the surfaces forward from the cockpit became heated, and a visor was used to deflect much of this heat from directly reaching the cockpit.[62]

Concorde had livery restrictions; the majority of the surface had to be covered with a highly reflective white paint to avoid overheating the aluminium structure due to heating effects from supersonic flight at Mach 2. In 1996, Air France briefly painted F-BTSD in a predominantly blue livery, with the exception of the wings, in a promotional deal with Pepsi.[63] In this paint scheme, Air France were advised to remain at Mach 2 for no more than 20 minutes at a time, but there was no restriction at speeds under Mach 1.7. F-BTSD was used because it was not scheduled for any long flights that required extended Mach 2 operations.[64]

Structural issues

Due to the high speeds at which Concorde travelled, large forces were applied to the aircraft's structure during banks and turns. This caused twisting and the distortion of the aircraft's structure. In addition there were concerns over maintaining precise control at supersonic speeds; both of these issues were resolved by active ratio changes between the inboard and outboard elevons, varying at differing speeds including supersonic. Only the innermost elevons, which are attached to the stiffest area of the wings, were active at high speed.[65] Additionally, the narrow fuselage meant that the aircraft flexed.[36] This was visible from the rear passengers' viewpoints.[66]

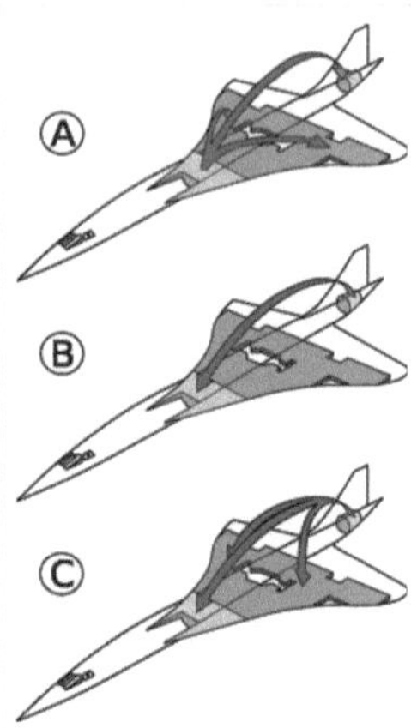

Fuel pitch trim

When any aircraft passes the critical mach of that particular airframe, the centre of pressure shifts rearwards. This causes a pitch down force on the aircraft if the centre of mass remains where it was. The engineers designed the wings in a specific manner to reduce this shift, but there was still a shift of about 2 metres. This could have been countered by the use of trim controls, but at such high speeds this would have caused a dramatic increase in the drag on the aircraft. Instead, the distribution of fuel along the aircraft was shifted during acceleration and deceleration to move the centre of mass, effectively acting as an auxiliary trim control.[67]

Range

In order to fly non-stop across the Atlantic Ocean, Concorde was developed to have the greatest supersonic range of any aircraft. This was achieved by a combination of engines which were highly efficient at supersonic speeds,[68] [36] a slender fuselage with high fineness ratio, and a complex wing shape for a high lift to drag ratio. This also required carrying only a modest payload and a high fuel capacity, and the aircraft was trimmed with precision to avoid unnecessary drag.[7] [67]

Nevertheless, soon after Concorde began flying, a Concorde "B" model was designed with slightly larger fuel capacity and slightly larger wings with leading edge slats to improve aerodynamic performance at all speeds, with the objective of expanding the range to reach markets in new regions.[69] It featured more powerful engines with sound deadening and without the fuel-hungry and noisy reheat. It was speculated that it was reasonably possible to create an engine with up to 25% gain in efficiency over the Rolls-Royce/Snecma Olympus 593.[70] This would have given 500 mi (805 km) additional range and a greater payload, making new commercial routes possible. This was cancelled due in part to poor sales of Concorde, but also to the rising cost of aviation fuel in the 1970s.[71]

Cabin pressurisation and radiation concerns

External view of Concorde's fuselage

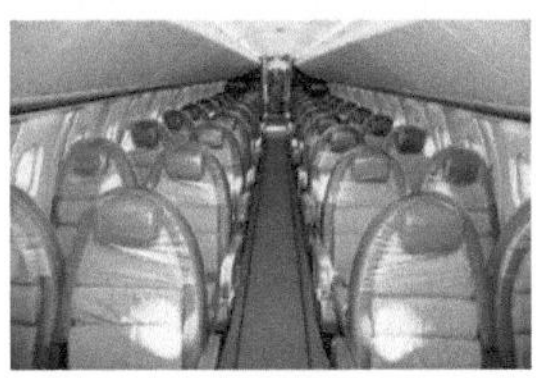

British Airways Concorde interior. The narrow fuselage permitted only 4 seats across the aircraft with limited headroom

The high altitude at which Concorde cruised meant passengers received almost twice the flux of extraterrestrial ionising radiation as those travelling on a conventional long-haul flight.[72] [73] Upon Concorde's introduction, it was speculated that this exposure during supersonic travels would increase the likelihood of skin cancer.[74] Due to the proportionally reduced flight time, the overall equivalent dose would normally be less than a conventional flight over the same distance.[75] Unusual solar activity might lead to an increase in incident radiation.[76] To prevent incidents of excessive radiation exposure, the flight deck had a radiometer and an instrument to measure the rate of decrease of radiation.[73] If the radiation level became too high, Concorde would descend below 47000 feet (m).

Airliner cabins were usually maintained at a pressure equivalent to 6,000–8,000 feet (1,800–2,400 m) elevation. Concorde's pressurisation was set to an altitude at the lower end of this range, 6000 feet (m).[77] Concorde's maximum cruising altitude was 60000 feet (m); subsonic airliners typically cruise below 40000 feet (m). Above 50000 feet (m), the lack of air pressure would give a "time of useful consciousness" in even a conditioned athlete of no more than 10–15 seconds.[78]

A sudden reduction in cabin pressure is hazardous to all passengers and crew.[79] At Concorde's altitude, the air density is very low; a breach of cabin integrity would result in a loss of pressure severe enough so that the plastic emergency oxygen masks installed on other passenger jets would not be effective, and passengers would quickly suffer from hypoxia despite quickly donning them. Concorde was equipped with smaller windows to reduce the rate of loss in the event of a breach,[80] a reserve air supply system to augment cabin air pressure, and a rapid descent procedure to bring the aircraft to a safe altitude. The FAA enforces minimum emergency descent rates for aircraft and made note of Concorde's higher operating altitude, concluding that the best response to a loss of pressure would be a rapid descent.[81] Continuous positive airway pressure would have delivered pressurised oxygen directly to the pilots through masks.[80]

Flight characteristics

Concorde performing a low-level flypast at an air show in August 1981

While commercial jets take eight hours to fly from New York to Paris, the average supersonic flight time on the transatlantic routes was just under 3.5 hours. Concorde had a maximum cruise altitude of **unknown operator: u','** metres (**unknown operator: u'strong'** ft) and an average cruise speed of Mach 2.02, about 1155 knots (2140 km/h or 1334 mph), more than twice the speed of conventional aircraft.[82]

With no other civil traffic operating at its cruising altitude of about 56000 ft (m), dedicated oceanic airways or "tracks" were used by Concorde to cross the Atlantic. Due to the nature of high altitude winds, these SST tracks were fixed in terms of their co-ordinates, unlike the North Atlantic Tracks at lower altitudes whose co-ordinates alter daily according to forecast weather patterns.[83] Concorde would also be cleared in a 15000-foot (m) block, allowing for a slow climb from 45,000 to 60000 ft (m) during the oceanic crossing as the fuel load gradually decreased.[84] In regular service, Concorde employed an efficient *cruise-climb* flight profile following take-off.[85]

The delta-shaped wings forced Concorde to attain a higher angle of attack at low speeds than conventional aircraft, but it allowed the formation of large low pressure vortices over the entire upper wing surface, maintaining lift.[86] The normal landing speed was 170 miles per hour (274 km/h).[87] Because of this high angle, during a landing approach Concorde was on the "back side" of the drag force curve, where raising the nose would increase the sink rate; the aircraft was thus largely flown on the throttle and was fitted with an autothrottle to reduce the pilot's workload.[88]

Brakes and undercarriage

Concorde rear undercarriage

Tail bumper of Concorde G-BOAG at the Museum of Flight in Seattle

Because of the way Concorde's delta-wing generated lift, the undercarriage had to be unusually strong. At rotation, Concorde would rise to a high angle of attack, about 18 degrees. Prior to rotation the wing generated almost no lift, unlike typical aircraft wings. Combined with the high airspeed at rotation (199 knots indicated airspeed), this increased the stresses on the rear undercarriage in a way that was initially unexpected during the development and required a major redesign.[89] Due to the high angle needed at rotation, a small set of wheels were added aft to prevent tailstrikes. The rear main undercarriage units swing towards each other to be stowed but due to their great height also need to retract telescopically before swinging in order to clear each other when stowed.[90]

Additionally, due to the high average takeoff speed of 250 miles per hour (km/h), Concorde needed upgraded brakes. Like most airliners, Concorde has anti-skid braking – a system which prevents the tyres from losing traction when the brakes are applied for greater control during roll-out. The brakes, developed by Dunlop, were the first carbon-based brakes used on an airliner.[91] The use of carbon over equivalent steel brakes provided a weight-saving of 12000 lb (5443 kg).[92]

Droop nose

Concorde's drooping nose, developed by Marshall Aerospace,[93] enabled the aircraft to switch between being streamlined to reduce drag and achieve optimum aerodynamic efficiency, and not obstructing the pilot's view during taxi, takeoff, and landing operations. Due to the high angle of attack the long pointed nose obstructed the view and necessitated the capability to droop. The droop nose was accompanied by a moving visor that retracted into the nose prior to being lowered. When the nose was raised to horizontal, the visor would raise in front of the cockpit windscreen for aerodynamic streamlining.[93]

Concorde landing at Farnborough in September 1974.

A controller in the cockpit allowed the visor to be retracted and the nose to be lowered to 5° below the standard horizontal position for taxiing and takeoff. Following takeoff and after clearing the airport, the nose and visor were raised. Prior to landing, the visor was again retracted and the nose lowered to 12.5° below horizontal for maximum visibility. Upon landing the nose was raised to the five-degree position to avoid the possibility of damage.[93]

The Federal Aviation Administration had objected to the restrictive visibility of the visor used on the first two prototype Concordes and thus requiring alteration before the FAA would permit Concorde to serve US airports; this led to the redesigned visor used on the production and the four pre-production aircraft (101, 102, 201, and 202).[94] The nose window and visor glass needed to endure temperatures in excess of 100°C at supersonic flight were developed by Triplex.[95]

Operational history

Scheduled flights

Scheduled flights began on 21 January 1976 on the London–Bahrain and Paris–Rio (via Dakar) routes,[96] with BA flights using the "*Speedbird Concorde*" call sign to notify air traffic control of the aircraft's unique abilities and restrictions, but the French using their normal callsigns.[97] The Paris-Caracas route (via Azores) began on 10 April. The US Congress had just banned Concorde landings in the US, mainly due to citizen protest over sonic booms, preventing launch on the coveted transatlantic routes. The US Secretary of Transportation, William Coleman, gave permission for Concorde service to Washington Dulles International Airport, and Air France and British Airways simultaneously began service to Dulles on 24 May 1976.[98]

The official handover ceremony to British Airways of its first Concorde occurred on 15 January 1976 at Heathrow Airport

When the US ban on JFK Concorde operations was lifted in February 1977, New York banned Concorde locally. The ban came to an end on 17 October 1977 when the Supreme Court of the United States declined to overturn a lower court's ruling rejecting efforts by the Port Authority and a grass-roots campaign led by Carol Berman to continue the ban.[99] In spite of complaints about noise, the noise report noted that Air Force One, at the time a Boeing VC-137, was louder than Concorde at subsonic speeds and during takeoff and landing.[100] Scheduled service from Paris and London to New York's John F. Kennedy Airport began on 22 November 1977.[101]

Concorde in Singapore Airlines livery at Heathrow in 1979

In 1977, British Airways and Singapore Airlines shared a Concorde for flights between London and Singapore International Airport at Paya Lebar via Bahrain. The aircraft, BA's Concorde G-BOAD, was painted in Singapore Airlines livery on the port side and British Airways livery on the starboard side.[102] [103] The service was discontinued after three return flights because of noise complaints from the Malaysian government;[104] it could only be reinstated on a new route bypassing Malaysian airspace in 1979. A dispute with India prevented Concorde from reaching supersonic speeds in Indian airspace, so the route was eventually declared not viable and discontinued in 1980.[105]

Air France Concorde in 1977

During the Mexican oil boom, Air France flew Concorde twice weekly to Mexico City's Benito Juárez International Airport via Washington, DC, or New York City, from September 1978 to November 1982.[106] [107] The worldwide economic crisis during that period resulted in this route's cancellation; the last flights were almost empty. The routing between Washington or New York and Mexico City included a deceleration, from Mach 2.02 to Mach 0.95, to cross Florida subsonically and avoid creating a sonic boom over the state; Concorde then re-accelerated back to high speed while crossing the Gulf of Mexico. On 1 April 1989, on an around-the-world luxury tour charter, British Airways implemented changes to this routing that allowed G-BOAF to maintain Mach 2.02 by passing around Florida to the east and south. Periodically Concorde visited the region on similar chartered flights to Mexico City and Acapulco.[108]

From 1978 to 1980, Braniff International Airways leased 10 Concordes, five each from Air France and British Airways.[109] These were used on subsonic flights between Dallas-Fort Worth and Washington Dulles International

Airport, flown by Braniff flight crews.[110] Air France and British Airways crews then took over for the continuing supersonic flights to London and Paris.[111] The aircraft were registered in both the United States and their home countries; the European registration was covered while being operated by Braniff, retaining full AF/BA liveries. The flights were not profitable and typically less than 50% booked, forcing Braniff to end its tenure as the only US Concorde operator in May 1980.[112] [113]

British Caledonian interest

Following the launch of British Airways' Concorde services, Britain's other major airline, British Caledonian (BCal), set up a task force headed by Gordon Davidson, BA's former Concorde director, to investigate the possibility of their own Concorde operations.[114] [115] [116] This was seen as particularly viable for the airline's long-haul network as there were two unsold aircraft then available for purchase.[117] [118] [119]

One important reason for BCal's interest in Concorde was that the British Government's 1976 aviation policy review had opened the possibility of BA setting up supersonic services in competition with BCal's established sphere of influence. To counteract this potential threat, BCal considered their own independent Concorde plans, as well as a partnership with BA.[120] [121] BCal were considered most likely to have set up a Concorde service on the Gatwick–Lagos route, a major source of revenue and profits within BCal's scheduled route network;[122] [123] BCal's Concorde task force did assess the viability of a daily supersonic service complementing the existing subsonic widebody service on this route.[118] [121] [124]

BCal entered into a bid to acquire at least one Concorde.[117] [119] [124] However, BCal eventually arranged for two aircraft to be leased from BA and Aérospatiale respectively, to be maintained by either BA or Air France. BCal's envisaged two-Concorde fleet would have required a high level of aircraft utilisation to be cost-effective; therefore, BCal had decided to operate the second aircraft on a supersonic service between Gatwick and Atlanta, with a stopover at either Gander or Halifax.[118] Consideration was given to services to Houston and various points on its South American network at a later stage.[124] [125] Both supersonic services were to be launched at some point during 1980; however, steeply rising oil prices caused by the 1979 energy crisis led to BCal shelving their supersonic ambitions.[121]

British Airways buys its Concordes outright

By around 1981 in the UK, the future for Concorde looked bleak. The British government had lost money operating Concorde every year, and moves were afoot to cancel the service entirely. A cost projection came back with greatly reduced metallurgical testing costs because the test rig for the wings had built up enough data to last for 30 years and could be shut down. Despite this, the government was not keen to continue. In 1983, BA's managing director, Sir John King, convinced the government to sell the aircraft outright to British Airways for £16.5 million plus the first year's profits.[126] [127]

An Air France Concorde at John F. Kennedy International Airport in 1987

King recognised that, in Concorde, BA had a premier product that was underpriced. Market research had revealed that many customers thought Concorde was more expensive than it actually was; thus ticket prices were progressively raised to match these perceptions.[36] It is reported that British Airways then ran Concorde at a profit, unlike their French counterpart.[128] [129]

Between 1984 and 1991, British Airways flew a thrice-weekly Concorde service between London and Miami, stopping at Washington Dulles International Airport.[130] [131] Until 2003, Air France and British Airways continued to operate the New York services daily. Concorde routinely flew to Grantley Adams International Airport, Barbados, during the winter holiday season.[132]

Prior to the Air France Paris crash, several UK and French tour operators operated charter flights to European destinations on a regular basis;[133] [134] the charter business was viewed as lucrative by British Airways and Air France.[135]

Concorde Flight 4590 crash

On 25 July 2000, Air France Flight 4590, registration F-BTSC, crashed in Gonesse, France after departing from Paris Charles de Gaulle en route to John F. Kennedy International Airport in New York City, killing all 100 passengers and nine crew members on board the flight, and four people on the ground. It was the only fatal accident involving Concorde.

According to the official investigation conducted by the *Bureau d'Enquêtes et d'Analyses pour la Sécurité de l'Aviation Civile* (BEA), the crash was caused by a titanium strip that fell from a Continental Airlines DC-10 that had taken off minutes earlier. This metal fragment punctured a tyre on Concorde's left main wheel bogie during takeoff. The tyre exploded, a piece of rubber hit the fuel tank, and while the fuel tank was not punctured, the impact caused a shock-wave which caused one of the fuel valves in the wing to burst open. This caused a major fuel leak from the tank, which then ignited due to sparking electrical landing gear wiring severed by another piece of the same tyre. The crew shut down engine number 2 in response to a fire warning, and with engine number 1 surging and producing little power, the aircraft was unable to gain height or speed. The aircraft entered a rapid pitch-up then a violent descent, rolling left and crashing tail-low into the Hôtelissimo Les Relais Bleus Hotel in Gonesse.[136] On 6 December 2010, Continental Airlines and John Taylor, one of their mechanics, were found guilty of involuntary manslaughter,[137] but on 30 November 2012 a French court overturned the conviction, saying mistakes by Continental and Taylor did not make them criminally responsible.[138]

Prior to the accident, Concorde had been arguably the safest operational passenger airliner in the world in terms of passenger deaths-per-kilometres travelled with zero, but with a history of tyre explosions 60 times higher than subsonic jets.[139] Safety improvements were made in the wake of the crash, including more secure electrical controls, Kevlar lining to the fuel tanks and specially developed burst-resistant tyres.[140]

The first flight after the modifications departed from London Heathrow on 17 July 2001, piloted by BA Chief Concorde Pilot Mike Bannister. During the 3-hour 20-minute flight over the mid-Atlantic towards Iceland, Bannister attained Mach 2.02 and 60000 ft (m) before returning to RAF Brize Norton. The test flight, intended to resemble the London–New York route, was declared a success and was watched on live TV, and by crowds on the ground at both locations.[141] The first flight with passengers after the accident took place on 11 September 2001, which landed shortly before the World Trade Center attacks in the United States. This was not a revenue flight, as all the passengers were BA employees.[142]

Normal commercial operations resumed on 7 November 2001 by BA and AF (aircraft G-BOAE and F-BTSD), with service to New York JFK, where passengers were welcomed by then mayor Rudy Giuliani.[143] [144]

Retirement

Concorde G-BOAB in storage at London Heathrow Airport. This aircraft flew for 22,296 hours between its first flight in 1976 and its final flight in 2000.

On 10 April 2003, Air France and British Airways simultaneously announced that they would retire Concorde later that year.[145] They cited low passenger numbers following the 25 July 2000 crash, the slump in air travel following 11 September 2001, and rising maintenance costs. Although Concorde was technologically advanced when introduced in the 1970s, 30 years later its analogue cockpit was dated. There had been little commercial pressure to upgrade Concorde due to a lack of competing aircraft, unlike other airliners of the same era such as the Boeing 747.[146] By its retirement, it was the last aircraft in British Airways' fleet that had a flight engineer; other aircraft, such as the modernised 747-400, had eliminated the role.[147]

On the same day, Sir Richard Branson offered to buy British Airways' Concorde fleet at their "original price of £1" for service with Virgin Atlantic Airways. Branson claimed this to be the same token price that British Airways had paid the British government; however, BA denied this and refused the offer.[148] Branson wrote in *The Economist* (23 October 2003) that his final offer was "over £5 million" and that he had intended to operate the fleet "for many years to come".[149] The chances for keeping Concorde in service were stifled by Airbus's lack of support for continued maintenance.[150] [151] [152]

It has been suggested that Concorde was not withdrawn for the reasons usually given but that it became apparent during the grounding of Concorde that the airlines could make more profit carrying first class passengers subsonically.[153] Rob Lewis suggested that the Air France retirement of its Concorde fleet was the result of a conspiracy between Air France Chairman Jean-Cyril Spinetta and Airbus CEO Noel Forgeard, and stemmed as much from a fear of being found criminally liable under French law for future Concorde accidents as from simple economics.[154] A lack of commitment to Concorde from Director of Engineering Alan MacDonald was cited as having undermined BA's resolve to continue operating Concorde.[155]

Promotional efforts in 1997 included a British Airways lottery to fly to New York held for 190 tickets valued at £5,400 each, to be offered up at £10. Callers had to ring a special hotline competing with up to 20 million people. The offer marked the 10th anniversary of BA's move into the private sector. Among the winners were 15 Scottish residents, including Joe Traynor and wife Margaret. Other winners included a health service worker from Glasgow who only asked for one ticket saying he couldn't think of anyone to take with him. Health Service Worker a Murali Murugasu, 28, said: "I didn't have time to tell anyone or ring them up so I will have to spend the time on my own. I am doing it just for the sheer experience of going on Concorde. At the time, British Airways were baffled why so many Scots got lucky. A spokeswoman said: "It was a good proportion, considering it was a worldwide promotion.[156]

Air France

Air France made its final commercial Concorde landing in the United States in New York City from Paris on 30 May 2003.[157] [158] Air France's final Concorde flight took place on 27 June 2003 when F-BVFC retired to Toulouse.[159]

Air France Concorde at Paris-Charles de Gaulle Airport

An auction of Concorde parts and memorabilia for Air France was held at Christie's in Paris on 15 November 2003; 1,300 people attended, and several lots exceeded their predicted values.[160] French Concorde F-BVFC was retired to Toulouse and kept functional for a short time after the end of service, in case taxi runs were required in support of the French judicial enquiry into the 2000 crash.[161] The aircraft is now fully retired and no longer functional.[162]

French Concorde F-BTSD has been retired to the "Musée de l'Air et de l'Espace" at Le Bourget (near Paris) and, unlike the other museum Concordes, a few of the systems are being kept functional. For instance, the famous "droop nose" can still be lowered and raised. This led to rumours that they could be prepared for future flights for special occasions.[163]

French Concorde F-BVFB currently rests at the Auto & Technik Museum Sinsheim at Sinsheim, Germany, after its last flight from Paris to Baden-Baden, followed by a spectacular transport to Sinsheim via barge and road. The museum also has a Tu-144 on display – this is the only place where both supersonic airliners can be seen together.[164]

British Airways

Mike Bannister (left) in the cockpit of BA002

British Airways conducted a North American farewell tour in October 2003. G-BOAG visited Toronto Pearson International Airport on 1 October, after which it flew to New York's John F. Kennedy International Airport.[165] G-BOAD visited Boston's Logan International Airport on 8 October, and G-BOAG visited Washington Dulles International Airport on 14 October.[166] It has been claimed that G-BOAD's flight from London Heathrow to Boston set a transatlantic flight record of 3 hours, 5 minutes, 34 seconds.[167] However the fastest transatlantic flight was from Heathrow to New York JFK airport on 7 February 1996, taking only 2 hours, 52 minutes, 59 seconds.[168] [169]

In a week of farewell flights around the United Kingdom, Concorde visited Birmingham on 20 October, Belfast on 21 October, Manchester on 22 October, Cardiff on 23 October, and Edinburgh on 24 October. Each day the aircraft made a return flight out and back into Heathrow to the cities, often overflying them at low altitude.[170] [171] [172] On 22 October, both Concorde flight BA9021C, a special from Manchester, and BA002 from New York landed simultaneously on both of Heathrow's runways. On 23 October 2003, the Queen consented to the illumination of Windsor Castle, an honour reserved for state events and visiting dignitaries, as Concorde's last west-bound commercial flight departed London.[173]

Concorde G-BOAC in her hangar at Manchester Airport Aviation Viewing Park

British Airways retired its Concorde fleet on 24 October.[174] G-BOAG left New York to a fanfare similar to that given for Air France's F-BTSD, while two more made round trips, G-BOAF over the Bay of Biscay, carrying VIP guests including former Concorde pilots, and G-BOAE to Edinburgh. The three aircraft then circled over London, having received special permission to fly at low altitude, before landing in sequence at Heathrow. The captain of the New York to London flight was Mike Bannister.[175]

The final flight of a Concorde in the US occurred on 5 November 2003 when G-BOAG flew from New York's Kennedy Airport to Seattle's Boeing Field to join the Museum of Flight's permanent collection. The plane was piloted by Mike Bannister and Les Broadie who claimed a flight time of three hours, 55 minutes and 12 seconds, a record between the two cities.[176] The museum had been pursuing a Concorde for their collection since 1984.[177] The final flight of a Concorde world-wide took place on 26 November 2003 with a landing at Filton, Bristol, UK.[178]

All of BA's Concorde fleet have been grounded, drained of hydraulic fluid and their airworthiness certificates withdrawn. Jock Lowe, ex-chief Concorde pilot and manager of the fleet estimated in 2004 that it would cost £10–15 million to make G-BOAF airworthy again.[163] BA maintain ownership and have stated that they will not fly again due to a lack of support from Airbus.[179] On 1 December 2003, Bonhams held an auction of British Airways' Concorde artifacts, including a nose cone, at Kensington Olympia in London.[180] [181] Proceeds of around £750,000 were raised, with the majority going to charity. G-BOAD is currently on display at the Intrepid Sea, Air & Space Museum in New York.[169] In 2007, BA announced that the advertising spot at Heathrow where a 40% scale model of Concorde was located would not be retained; the model is now on display at the Brooklands Museum.[182]

Restoration

Although only used for spares after being retired from test flying and trials work in 1981, Concorde G-BBDG was dismantled and transported by road from Filton then restored from essentially a shell at the Brooklands Museum in Surrey.[183]

One of the youngest Concordes (F-BTSD) is on display at Le Bourget Air and Space Museum in Paris. In February 2010, it was announced that the museum and a group of volunteer Air France technicians intend to restore F-BTSD so it can taxi under its own power.[184] On 29 May 2010, it was reported that a group comprising the British Save Concorde Group and the French Olympus 593 had begun work on inspecting the engines of a Concorde at Le Bourget Air and Space Museum, with the intent to restore the plane to be able to fly again in demonstrations and air shows.[185]

Comparable aircraft

The only supersonic airliner in direct competition with Concorde was the Soviet Tupolev Tu-144, nicknamed "Concordski" by Western Europeans for its outward similarity to Concorde.[186] It had been alleged that Soviet espionage efforts had resulted in the theft of Concorde blueprints, ostensibly to assist in the design of the Tu-144.[187] As a result of a rushed development programme, the first Tu-144 prototype was substantially different from the preproduction machines, but both were cruder than Concorde. The Tu-144*S* had a significantly shorter range than Concorde, due to its low-bypass turbofan engines.[188] The aircraft had poor control at low speeds because of a simpler supersonic wing design; in addition the Tu-144 required parachutes to land while Concorde used anti-lock brakes.[189] The Tu-144 had two crashes, one at the 1973 Paris Air Show,[190] [191] and another during a pre-delivery test flight in May 1978.[192] [193]

Later production Tu-144 versions were more refined and competitive. They had retractable canards for better low-speed control, turbojet engines providing nearly the fuel efficiency and range of Concorde[194] and a top speed of Mach 2.35. Passenger service commenced in November 1977, but after the 1978 crash the aircraft was taken out of service. The aircraft had an inherently unsafe structural design as a consequence of an automated production method chosen in order to simplify and speed up manufacturing.[195]

The American designs, the Boeing 2707 and the Lockheed L-2000 were to have been larger, with seating for up to 300 people.[196] [197] Running a few years behind Concorde, the Boeing 2707 was redesigned to a cropped delta layout; the extra cost of these changes helped to kill the project.[198] The operation of US military aircraft such as the XB-70 Valkyrie and B-58 Hustler had shown that sonic booms were quite capable of reaching the ground,[199] and the experience from the Oklahoma City sonic boom tests led to the same environmental concerns that hindered the commercial success of Concorde. The American government cancelled the project in 1971, after having spent more than $1 billion.[200]

The only other large supersonic aircraft comparable to Concorde are strategic bombers, principally the Russian Tupolev Tu-22, Tu-22M and Tu-160 and the American B-1 Lancer.[201]

Impact

Environmental

Prior to Concorde's flight trials, the developments made by the civil aviation industry were largely accepted by governments and their respective electorates. Opposition to Concorde's noise, particularly on the east coast of the United States,[202] [203] forged a new political agenda on both sides of the Atlantic, with scientists and technology experts across a multitude of industries beginning to take the environmental and social impact more seriously.[204] [205] Although Concorde led directly to the introduction of a general noise abatement programme for aircraft flying out of John F. Kennedy Airport, many found that Concorde was quieter than expected,[36] partly due to the pilots temporarily throttling back their engines to reduce noise during overflight of residential areas.[206] Even before the launch of revenue earning services, it had been noted that Concorde was quieter than several aircraft already commonly in service at that time.[207]

Concorde produced nitrogen oxides in its exhaust, which, despite complicated chemical interactions with other ozone-depleting chemicals, are understood to result in degradation to the ozone layer at the stratospheric altitudes it cruised.[208] It has been pointed out that other, lower-flying, airliners produce ozone during their flights in the troposphere, but vertical transit of gases between the layers is restricted. The small fleet operated meant overall ozone-layer degradation caused by Concorde was negligible.[208] David W. Fahey, of the National Oceanic and Atmospheric Administration, found that to produce a drop in stratospheric Ozone of 1 to 2% would require a fleet of 500 supersonic aircraft to be operated. Dr. Fahey stated that this would not be a limiting factor for further Supersonic Transport development.[209]

Concorde's technical leap forward boosted the public's understanding of conflicts between technology and the environment as well as awareness of the complex decision analysis processes that surround such conflicts.[210] In France, the use of acoustic fencing alongside TGV tracks might not have been achieved without the 1970s controversy over aircraft noise.[211] In the UK, the CPRE has issued tranquillity maps since 1990.[212]

Public perception

Parade flight at Queen's Golden Jubilee in June 2002

HM The Queen and HRH The Duke of Edinburgh disembark Concorde in 1991.

Concorde was normally perceived as a privilege of the rich, but special circular or one-way (with return by other flight or ship) charter flights were arranged to bring a trip within the means of moderately well-off enthusiasts.[213]

The aircraft was usually referred to by the British as simply "Concorde".[214] In France it was known as "le Concorde" due to "le", the definite article,[215] used in French grammar to introduce the name of a ship or aircraft,[216] and the capital being used to distinguish a proper name from a common noun of the same spelling.[215][217] In French, the common noun *concorde* means "agreement, harmony, or peace". [218] Concorde's pilots and British Airways in official publications often refer to Concorde both in the singular and plural as "she" or "her".[219][220] [221]

As a symbol of national pride, an example from the BA fleet made occasional flypasts at selected Royal events, major air shows and other special occasions, sometimes in formation with the Red Arrows.[222] [223] On the final day of commercial service, public interest was so great that grandstands were erected at Heathrow Airport. Significant numbers of people attended the final landings; the event received widespread media coverage.[224]

In 2006, 37 years after her first test flight, Concorde was announced the winner of the Great British Design Quest organised by the BBC and the Design Museum. A total of 212,000 votes were cast with Concorde beating design icons such as the Mini, mini skirt, Jaguar E-type, Tube map and the Supermarine Spitfire.[5]

Records

The fastest transatlantic airliner flight was from New York JFK to London Heathrow on 7 February 1996 by British Airways' G-BOAD in 2 hours, 52 minutes, 59 seconds from takeoff to touchdown.[168] Concorde also set other records, including the official FAI "Westbound Around the World" and "Eastbound Around the World" world air speed records.[225] On 12–13 October 1992, in commemoration of the 500th anniversary of Columbus' first New World landing, Concorde Spirit Tours (USA) chartered Air France Concorde F-BTSD and circumnavigated the world in 32 hours 49 minutes and 3 seconds, from Lisbon, Portugal, including six refuelling stops at Santo Domingo, Acapulco, Honolulu, Guam, Bangkok, and Bahrain.[226]

The eastbound record was set by the same Air France Concorde (F-BTSD) under charter to Concorde Spirit Tours in the USA on 15–16 August 1995. This promotional flight circumnavigated the world from New York/JFK International Airport in 31 hours 27 minutes 49 seconds, including six refuelling stops at Toulouse, Dubai, Bangkok, Andersen AFB in Guam, Honolulu, and Acapulco.[227] By its 30th flight anniversary on 2 March 1999 Concorde had clocked up 920,000 flight hours, with more than 600,000 supersonic, much more than all of the other supersonic aircraft in the Western world combined.[228]

On its way to the Museum of Flight in November 2003, G-BOAG set a New York City-to-Seattle speed record of 3 hours, 55 minutes, and 12 seconds.[229]

Replacements in development

The desire for a second-generation supersonic aircraft has remained within some elements of the aviation industry,[230] [231] and several concepts emerged quickly following the retirement of Concorde.

In November 2003, EADS—the parent company of Airbus—announced that it was considering working with Japanese companies to develop a larger, faster replacement for Concorde.[232] [233] In October 2005, JAXA, the Japan Aerospace eXploration Agency, undertook aerodynamic testing of a scale model of an airliner designed to carry 300 passengers at Mach 2 (working name *NEXST*). If pursued to commercial deployment, it would be expected to be in service around 2020–2025.[234] On 18 June 2011, the *Zero Emission High Speed Transport* or *ZEHST* concept aircraft was unveiled by EADS at the Paris Air Show.[235] The ZEHST, a hypersonic aircraft to be capable of 3,000 mph (4,800 km/h), is a result of the collaboration efforts between EADS and Japan.[236]

The British company Reaction Engines Limited, with 50% EU money, has been engaged in a research programme called *LAPCAT*, which examined a design for a hydrogen-fuelled plane carrying 300 passengers called the *A2*, potentially capable of flying at Mach 5+ non-stop from Brussels to Sydney in 4.6 hours.[237] The follow-on research effort, *LAPCAT II* began in 2008 and is to last four years.[238]

In May 2008, it was reported that Aerion Corporation had $3 billion of pre-order sales on its Aerion SBJ supersonic business jet.[239] In late 2010, the project continued with a testbed flight of a section of the wing.[240]

Supersonic Aerospace International's Quiet Supersonic Transport was a 12 passenger design from Lockheed Martin that was to cruise at Mach 1.6, and was to have created a sonic boom only 1% as strong as that generated by Concorde.[241]

Operators

- Air France
- British Airways
- Braniff International Airways (short term lease)[112]
- Singapore Airlines (short term wet lease)[103]

Specifications

Data from Wall Street Journal,[147] *The Concorde Story,*[242] *The International Directory of Civil Aircraft,*[38] *Richard Seamen aircraft museum*[243]

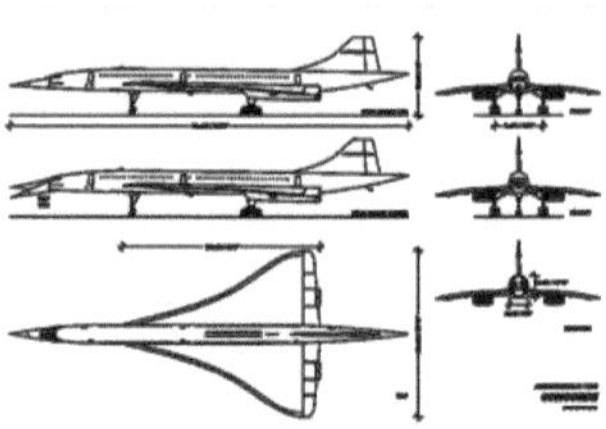

Other line drawings of Concorde

General characteristics

- **Crew:** 3 (2 Pilots and a flight engineer)
- **Capacity:** 92–120 passengers (128 in high-density layout)[244]
- **Length:** 202 ft 4 in (61.66 m)
- **Wingspan:** 84 ft 0 in (25.6 m)
- **Height:** 40 ft 0 in (12.2 m)
- **Fuselage internal length:** 129 ft 0 in (m)
- **Fuselage width:** maximum of 9 ft 5 in (m) external 8 ft 7 in (m) internal
- **Fuselage height:** maximum of 10 ft 10 in (m) external 6 ft 5 in (m) internal)
- **Wing area:** 3,856 ft^2 (358.25 m^2)

- **Empty weight:** 173,500 lb (78,700 kg)
- **Useful load:** 245,000 lb (111,130 kg)
- **Powerplant:** 4 × Rolls-Royce/SNECMA Olympus 593 Mk 610 afterburning turbojets
    - **Dry thrust:** 32,000 lbf (140 kN) each
    - **Thrust with afterburner:** 38,050 lbf (169 kN) each
- **Maximum fuel load:** 210940 lb (kg)
- **Maximum taxiing weight:** 412000 lb (kg)

Performance

Concorde G-BOAC

- **Maximum speed:** Mach 2.04 (≈1,354 mph, 2,179 km/h) at cruise altitude
- **Cruise speed:** Mach 2.02 (≈1,340 mph, 2,158 km/h) at cruise altitude
- **Range:** 3,900 nmi (4,500 mi, 7,250 km)
- **Service ceiling:** 60,000 ft (18,300 m)
- **Rate of climb:** 5,000 ft/min (25.41 m/s)
- **lift-to-drag:** *Low speed*– 3.94, *Approach*– 4.35, *250 kn, 10,000 ft*– 9.27, *Mach 0.94*– 11.47, *Mach 2.04*– 7.14
- **Fuel consumption:** 46.85 lb/mi (13.2 kg/km) operating for maximum range
- **Thrust/weight:** 0.373
- **Maximum nose tip temperature:** 260 °F (°C)

See also

- BAC 221
- Barbara Harmer, the first qualified female Concorde pilot.
- North American XB-70 Valkyrie
- Richard Wiggs, Anti-Concorde campaigner

Related development

- Bristol Type 223
- Sud Aviation Super-Caravelle
- Fairey Delta

Related lists

- List of civil aircraft

References

Notes

[1] Towey 2007, p. 359.

[2] "Ageing luxury jet" (http://news.bbc.co.uk/1/hi/uk/850899.stm). *BBC News*, 25 July 2000. 25 July 2000. .

[3] Marston, Paul (16 August 2000). "Is this the end of the Concorde dream?" (http://www.telegraph.co.uk/news/uknews/1352927/Is-this-the-end-of-the-Concorde-dream.html). London: *The Daily Telegraph*, 16 August 2000. .

[4] http://news.bbc.co.uk/2/hi/uk_news/2934257.stm

[5] Jury, Louise (17 March 2006). "Concorde beats Tube map to become Britain's favourite design" (http://news.independent.co.uk/uk/this_britain/article351851.ece). London: *The Independent*, 17 March 2006. .

[6] "Early History." (http://www.concordesst.com/history/eh1.html#a) *concordesst.com*. Retrieved 8 September 2007.

[7] Maltby, R.L.. "The development of the slender delta concept" (http://md1.csa.com/partners/viewrecord.php?requester=gs&collection=TRD&recid=A6825061AH&q=Concorde+Delta&uid=788872723&setcookie=yes). *Aircraft Engineering* **40, 1968**. .

[8] "Aerospace: Pan Am's Concorde Retreat" (http://www.time.com/time/magazine/article/0,9171,903850,00.html). *Time*, 12 February 1973. 12 February 1973. .

[9] "Vertrag mit Luken" (http://www.spiegel.de/spiegel/print/d-46437753.html). Der Spiegel. 13 March 1967. . Retrieved 6 November 2012.

[10] Taylor 1965, p. 130.

[11] Winchester 2005, p. 134.

[12] Benn, Tony (17 October 2003). "Sonic booms and that 'e' on the end: Tony Benn remembers his role in getting Concorde off the ground" (http://www.guardian.co.uk/g2/story/0,3604,1064752,00.html). London: *The Guardian,* 17 October 2003. .

[13] McIntyre 1992, p. 20.

[14] Note this British convention is used throughout this article: "In depth: Farewell to Concorde" (http://news.bbc.co.uk/2/hi/in_depth/uk/2003/concorde_retirement/default.stm). *BBC News,* 15 August 2007. 15 August 2007. .

[15] "About Concorde – main page" (http://www.britishairways.com/concorde/aboutconcorde.html). *British Airways*. . Retrieved 11 January 2010.

[16] "Pilot Says Concorde Flight "Perfect"" (http://news.google.co.uk/newspapers?id=KXoyAAAAIBAJ&sjid=GLkFAAAAIBAJ&pg=6573,360672&dq=concorde&hl=en). *Montreal Gazette,*. 1969. . Retrieved 30 June 2011.

[17] "Concorde Tops Speed of Sound for 9 Minutes on a Test Flight" (http://select.nytimes.com/gst/abstract.html?res=F30C1EFC3454127B93C0A9178BD95F4D8685F9). *The New York Times,* 2 October 1969. 2 October 1969. .

[18] "1969: Concorde flies for the first time" (http://news.bbc.co.uk/onthisday/hi/dates/stories/march/2/newsid_2514000/2514535.stm). *BBC News,* 2 March 1969. 2 March 1969. . Retrieved 8 July 2007.

[19] Rohrbach, Edward (10 April 1969). "Concorde 002 Makes 1st Flight" (http://pqasb.pqarchiver.com/chicagotribune/access/584606362.html?dids=584606362:584606362&FMT=ABS&FMTS=ABS:AI&type=historic&date=Apr+10,+1969&author=&pub=Chicago+Tribune&desc=Concorde+002+Makes+1st+Flight&pqatl=google). *Chicago Tribune,* 10 April 1969. . Retrieved 30 June 2011.

[20] "Concorde 001 Makes Its First Atlantic Crossing" (http://pqasb.pqarchiver.com/chicagotribune/access/597531692.html?dids=597531692:597531692&FMT=CITE&FMTS=CITE:AI&type=historic&date=Sep+05,+1971&author=&pub=Chicago+Tribune&desc=Concorde+001+Makes+Its+First+Atlantic+Crossing&pqatl=google). *Chicago Tribune,* 5 September 1971. 5 September 1971. . Retrieved 30 June 2011.

[21] "Anglo-French Concorde Lands in Brazil to begin Week of Demonstration Flights" (http://news.google.co.uk/newspapers?id=VCY0AAAAIBAJ&sjid=2uAIAAAAIBAJ&pg=2521,1962837&dq=concorde&hl=en). Bangor Daily News. 7 September 1971. .

[22] Stern, Michael (3 June 1972). "Concorde Prototype Begins 10-Nation Tour; Britain Shows Optimism For Supersonic Aircraft" (http://select.nytimes.com/gst/abstract.html?res=F70D10F73B591A7493C1A9178DD85F468785F9). *The New York Times,* 3 June 1972. . Retrieved 30 June 2011.

[23] Witkin, Richard (21 September 1973). "A Supersonic Concorde Lands in Texas" (http://select.nytimes.com/gst/abstract.html?res=FA0E16FB3A54137A93C3AB1782D85F478785F9). *The New York Times,* 21 September 1973. . Retrieved 30 June 2011.

[24] "Concordes limited to 16". *Virgin Islands Daily News,* 5 June 1976.

[25] "Payments for Concorde" (http://www.britishairways.com/concorde/faq.html#6). *British Airways*. . Retrieved 2 December 2009.

[26] Lewis, Anthony (12 February 1973). "Britain and France have wasted billions on the Concorde" (http://select.nytimes.com/gst/abstract.html?res=F50D16F93C54137A93C0A81789D85F478785F9). *The New York Times,* 12 February 1973. .

[27] "Malaysia lifting ban on the use Of its Airspace by the Concorde" (http://select.nytimes.com/gst/abstract.html?res=F70813F83F5511728DDDAE0994DA415B888BF1D3). *The New York Times,* 17 December 1978. 17 December 1978. . Retrieved 30 June 2011.

[28] "News from around the world" (http://news.google.co.uk/newspapers?id=xTMsAAAAIBAJ&sjid=48sEAAAAIBAJ&pg=6855,2298265&dq=india+concorde&hl=en). *Herald-Journal,* 13 January 1978. . Retrieved 30 June 2011.

[29] Ross 1978, p. 46.

[30] Marston, Paul (16 August 2000). "Is this the end of the Concorde dream?" (http://www.telegraph.co.uk/news/uknews/1352927/Is-this-the-end-of-the-Concorde-dream.html). London: *Daily Telegraph,* 16 August 2000. .

[31] Ross 1978, pp. 47–49.

[32] Ross 1978, p. 49.

[33] Gunn 2010, p. 45.

[34] Favre, C. (1996). Mark B. Tischler. ed. *Advances in aircraft flight control* (http://books.google.com/?id=DbUhEnlI3OkC&pg=PA211&dq=wire+Concorde#v=onepage&q=wire Concorde&f=false). CRC Press. p. 219. ISBN 0-7484-0479-1. . Retrieved 30 June 2011.

[35] Masefield, Peter. "Obituary: Sir Archibald Russell". UK: *The Independent,* 1 July 1995.

[36] "NOVA transcript: Supersonic Dream" (http://www.pbs.org/wgbh/nova/transcripts/3203_concorde.html). *PBS,* 18 January 2005. .

[37] "Rolls-Royce Snecma Olympus" (http://www.janes.com/transport/news/jae/jae000725_1_n.shtml). Janes. 25 July 2000. .

[38] Frawley 2003, p. 14.

[39] "Unified propulsion 8" (http://ocw.mit.edu/ans7870/16/16.unified/propulsionS04/UnifiedPropulsion8/UnifiedPropulsion8.htm). MIT. . Retrieved 8 December 2010.

[40] "*Concorde* – Choice of a light alloy for the construction of the first supersonic commercial aircraft" (http://md1.csa.com/partners/viewrecord.php?requester=gs&collection=TRD&recid=A6424111AH&q=Concorde+aluminium+construction&uid=789267644&setcookie=yes). *Revue De L'Aluminium* (316, March 1964): 111–119. .

[41] Wolfe, B.S. (1967). "The Concorde Automatic Flight Control System: A description of the automatic flight control system for the Anglo/French SST and its development to date" (http://www.emeraldinsight.com/journals.htm?articleid=1681566&show=pdf). *Aircraft Engineering and Aerospace Technology* (MCB UP) **39** (5, 1967): 40. doi:10.1108/eb034268. ISSN 0002-2667. .

[42] Schefer, L.J.. "Concorde has designed-in reliability" (http://md1.csa.com/partners/viewrecord.php?requester=gs&collection=TRD&recid=A7716749AH&q=Concorde+hydraulic&uid=788858323&setcookie=yes). *Hydraulics and Pneumatics* **29, 1976**: 51–55. .

[43] Owen 2001, p. 101.

[44] "Aircraft Stopping Systems" (http://www.emeraldinsight.com/Insight/viewContentItem.do?contentType=Article&hdAction=lnkpdf&contentId=1682576). *Aircraft Engineering and Aerospace Technology* (MCB UP) **47** (10, 1975): 18. 1975. doi:10.1108/eb035278. ISSN 0002-2667. .

[45] Turner, H.G. (1971). "Fuel Management for Concorde: A brief account of the fuel system and the fuel pumps developed for the aircraft" (http://www.emeraldinsight.com/Insight/viewContentItem.do?contentType=Article&contentId=1682044). *Aircraft Engineering and Aerospace Technology* (MCB Ltd) **43** (3, 1971): 36–39. doi:10.1108/eb034738. ISSN 0002-2667. .

[46] "British Contribution to Concord Production in France" (http://www.emeraldinsight.com/Insight/viewContentItem.do?contentType=Article&hdAction=lnkpdf&contentId=1681209). *Aircraft Engineering and Aerospace Technology* (MCB Ltd) **36** (8, 1964): 232–237. 1964. doi:10.1108/eb033911. ISBN 0002-2667. .

[47] Owen 2001, p. 206.

[48] Birtles, Philip. *Concorde*, pp. 62–63. Vergennes, Vermont: Plymouth Press, 2000. ISBN 1-882663-44-6.

[49] "Rolls Royce Olympus history." (http://www.wingweb.co.uk/engines/Olympus593_MK610.html) *wingweb.co.uk*. Retrieved 15 January 2010.

[50] Ganley, G.; G. Laviec (1989). *The Rolls Royce/SNECMA Olympus 593 engine operational experience and the lessons learned* (http://md1.csa.com/partners/viewrecord.php?requester=gs&collection=TRD&recid=A9110958AH&q=Rolls-Royce/Snecma+Olympus+593&uid=788858323&setcookie=yes). *European Symposium on the Future of High Speed Air Transport*. pp. 73–80. . Retrieved 30 June 2011.

[51] Lynam, Joe (19 July 2006). "Are the skies turning green?" (http://news.bbc.co.uk/1/hi/business/5195964.stm). BBC News. .

[52] Ganley, G. A. (September 1991). *Concorde Propulsion—Did we get it right? The Rolls-Royce/Snecma Olympus 593 Engine reviewed* (https://www.sae.org/technical/papers/912180). SAE International. .

[53] Fisher, S.A.; M.C. Neale and A.J. Brooks. "On the sub-critical stability of variable ramp intakes at Mach numbers around 2" (http://naca.central.cranfield.ac.uk/reports/arc/rm/3711.pdf). *Aeronautical Research Council* (Her Majesty's Stationery Office, 1972). .

[54] "Concorde Special – The test pilot – John Cochrane" (http://www.flightglobal.com/articles/2003/10/21/172657/concorde-special-the-test-pilot-john-cochrane.html). Flight International. 21 October 2003. .

[55] Woodman, Peter (27 July 2000). "How a Concorde pilot would handle a nightmare failure" (http://www.highbeam.com/doc/1G1-63710463.html). Birmingham Post. .

[56] Dalton, Alastair. "Supersonic: The enduring allure of Concorde" (http://www.scotsman.com/news/transport/supersonic-the-enduring-allure-of-concorde-1-2415088). Scotsman.com, 17 July 2012.

[57] Eberhart, Jonathan. "When the SST Is Too Slow..". *Science News* (Society for Science & the Public) **91** (22, 3 June 1967): 528–529. doi:10.2307/3951418. JSTOR 3951418.

[58] "The Concorde takes shape : Test programme and construction proceeding according to schedule" (http://www.emeraldinsight.com/Insight/viewContentItem.do?contentType=Article&hdAction=lnkpdf&contentId=1681441). *Aircraft Engineering and Aerospace Technology* (MCB UP) **38** (4, 1966): 38. 1966. doi:10.1108/eb034143. ISSN 0002-2667. .

[59] N'guyen, V.P.; Perrais; J.P. Perrais (1972). "Fatigue Tests on Big Structure Assemblies of Concorde Aircraft". *Advanced Approaches to Fatigue Evaluation. NASA SP-309* (NASA) **309**: 631. Bibcode 1972NASSP.309..631N.

[60] Wallace, James (7 November 2003). "Those who flew the Concorde will miss it" (http://www.seattlepi.com/business/147276_pilot07.html). Seattle Post. .

[61] Gedge, G.T.; M.I. Prod. "Introduction to Concorde: A brief review of the Concorde and its prospects". *Aircraft Engineering and Aerospace Technology* (Emerald Group Publishing Limited) **40** (3, 1993).

[62] Owen 2001, p. 14.

[63] "Is this the colour of the new millennium?" (http://www.highbeam.com/doc/1P2-4794332.html). UK: *The Independent,* 3 April 1996. .

[64] Frade, Cristina. "Azul contra rojo". *El Mundo,* 5 April 1996.

[65] Owen 2001, p. 78.

[66] Kocivar, Ben. "Aboard the Concorde SST." (http://books.google.co.uk/books?id=lpiMSzja6W4C&lpg=PA142&dq=concorde flexing&pg=PA118#v=onepage&q=concorde flexing&f=false) *Popular Science,* October 1973, p. 117.

[67] "Flight Refuelling Limited and Concorde: The fuel system aboard is largely their work" (http://www.emeraldinsight.com/Insight/viewContentItem.do?contentType=Article&hdAction=lnkpdf&contentId=1682642). *Aircraft Engineering and Aerospace Technology* (MCB UP) **48** (9, September 1976): 20–21. doi:10.1108/eb035344. ISSN 0002-2667. .

[68] "It is the world's most efficient jet engine." -Ricky Bastin, Concorde Engineer

[69] "Concorde model 'B'" (http://www.concordesst.com/concordeb.html)

[70] Strack, William. *Propulsion challenges and opportunities for high-speed transport aircraft* (http://md1.csa.com/partners/viewrecord.php?requester=gs&collection=TRD&recid=N9222540AH&q=Concorde+engines&uid=788858323&setcookie=yes). *Aeropropulsion,* 1987. pp. 437–452. . Retrieved 30 June 2011.

[71] Smale, Alison. "Fuel costs kill Second Generation of Concordes" (http://news.google.co.uk/newspapers?id=Q-0cAAAAIBAJ&sjid=rGcEAAAAIBAJ&pg=6914,3256355&dq=tu+144&hl=en). *Sarasota Herald-Tribune,* 22 September 1979. .

[72] "How much radiation might I be exposed to?" (http://www.britishairways.com/travel/healthcosmic/public/en_gb#4). British Airways. . Retrieved 11 January 2010.

[73] Guerin, D.W. (1973). "Electronic safety test replaces radioactive test source" (http://www.emeraldinsight.com/Insight/viewContentItem.do;jsessionid=E0A02B0587619C9A40D5736FCE7B3F02?contentType=Article&hdAction=lnkpdf&contentId=1682309). *Aircraft Engineering and Aerospace Technology* (*MCB* UP) **45** (4, 1973): 10. doi:10.1108/eb035011. ISSN 0002-2667. .
[74] "Skin cancer danger linked to stratospheric jet planes" (http://news.google.co.uk/newspapers?id=GfUNAAAAIBAJ&sjid=dnkDAAAAIBAJ&pg=2948,4208&dq=second+generation+concorde&hl=en). St. Petersburg Times. 1 April 1975. .
[75] "Cosmic radiation" (http://www.britishairways.com/travel/healthcosmic/public/en_gb). British Airways. . Retrieved 11 January 2010.
[76] Arctowski, Henryk (1940). "On Solar Faculae and Solar Constant Variations" (http://www.pnas.org/cgi/reprint/26/6/406.pdf) (PDF). *Proceedings of the National Academy of Sciences* (*Proc. Natl. Acad. Sci. U.S.A.*, 1940) **26** (6): 406–411. Bibcode 1940PNAS...26..406A. doi:10.1073/pnas.26.6.406. PMC 1078196. PMID 16588370. .
[77] Hepburn, A.N. (1967). "Human Factors in the Concorde" (http://occmed.oxfordjournals.org/cgi/reprint/17/2/47.pdf). *Occupational Medicine* **17** (2, 1967): 47–51. doi:10.1093/occmed/17.2.47. .
[78] Wolff, Mark. "Cabin Decompression and Hypoxia" (http://www.theairlinepilots.com/medical/decompressionandhypoxia.htm). *PIA Air Safety Publication,*6 January 2006. .
[79] *Flight Training Handbook* (http://books.google.com/books?id=ioRTAAAAMAAJ). U.S. Dept. of Transportation, Federal Aviation Administration, Flight Standards Service, 1980. 1980. p. 250. . Retrieved 28 July 2007.
[80] Nunn 1993, p. 341.
[81] Happenny, Steve. "Interim Policy on High Altitude Cabin Decompression – Relevant Past Practice" (http://rgl.faa.gov/Regulatory_and_Guidance_Library\rgPolicy.nsf/0/90AA20C2F35901D98625713F0056B1B8?OpenDocument). *Federal Aviation Administration,* 24 March 2006. .
[82] Schrader 1989, p. 64.
[83] Orlebar 2002, p. 84.
[84] Prestwick Oceanic Area Control Centre: Manual of Air Traffic Services (Part 2). NATS
[85] Orlebar 2002, p. 92.
[86] Orlebar 2002, p. 44.
[87] Schrader 1989, p. 84.
[88] Orlebar 2002, p. 110.
[89] Rose, David. "The real story of Flight 4590: Special Investigation" (http://www.iasa.com.au/folders/Safety_Issues/others/concordespacer.html). *iasa.com.au,* 13 May 2001. .
[90] Brooklands Museum
[91] Stimson, I. L.; Fisher; R. Fisher. "Design and Engineering of Carbon Brakes". *Philosophical Transactions of the Royal Society of London* (The Royal Society) **294** (1411, January 1980): 583–590. Bibcode 1980RSPTA.294..583S. doi:10.1098/rsta.1980.0068. JSTOR 36383.
[92] Owen 2001, p. 118.
[93] "Droop nose" (http://www.flightglobal.com/pdfarchive/view/1971/1971 - 1503.html). Flight International. 12 August 1971. pp. 257–258. .
[94] Owen 2001, p. 84.
[95] "Triplex in Concorde: The story behind the film." (http://www.flightglobal.com/pdfarchive/view/1968/1968 - 2105.html) *Flightglobal.com,* 1968. Retrieved 7 June 2011.
[96] Strang, Dr. W.J; R. McKinley. "Concorde in Service" (http://www.emeraldinsight.com/Insight/viewContentItem.do;jsessionid=2594967949E5659C545A433375AEF406?contentType=Article&hdAction=lnkpdf&contentId=1682798). *Aircraft Engineering and Aerospace Technology* (MCB UP) **50** (12, 1978). doi:10.1108/eb035500. ISSN 0002-2667. .
[97] "BA Tribute to Concorde. The takeoff scene at the end of the video contains a clip of the ATC communication with the "Speedbird Concorde"" (http://video.google.com/videoplay?docid=7422409257931235094). British Airways. . Retrieved 11 January 2010.
[98] Donin, Robert B. *Safety Regulation of the Concorde Supersonic Transport: Realistic Confinement of the National Environmental Policy Act* (http://heinonline.org/HOL/LandingPage?collection=journals&handle=hein.journals/tportl8&div=7&id=&page=). *HeinOnline,* 1976. . Retrieved 30 June 2011.
[99] O'Grady, Jim (27 April 2003). "Neighborhood Report: The Rockaways; Ears Ringing? It's Cheering Over the Demise Of the Concorde" (http://query.nytimes.com/gst/fullpage.html?res=9906E2DC153DF934A15757C0A9659C8B63). *The New York Times,* 27 April 2003. . Retrieved 30 June 2011.
[100] "The Nation: Smooth Landing for the Birds" (http://www.time.com/time/magazine/article/0,9171,915764,00.html). *Time,* 5 December 1977. 5 December 1977. . Retrieved 30 June 2011.
[101] "Concorde facts and figures" (http://www.britishairways.com/concorde/aboutconcorde.html#facts_figures). *British Airways*. . Retrieved 11 January 2010.
[102] Warneke, Ross. "Concorde by June: Offer to Qantas" (http://news.google.co.uk/newspapers?id=3PkQAAAAIBAJ&sjid=hpIDAAAAIBAJ&pg=4864,7521914&dq=concorde+singapore+airlines&hl=en). Australia: *The Age,* 25 October 1977. .
[103] "Singapore Concorde flights" (http://select.nytimes.com/gst/abstract.html?res=F70611FE3C5E167493C6A8178BD95F438785F9). *The New York Times,* 14 October 1977. 14 October 1977. . Retrieved 30 June 2011.
[104] "London and Singapore halt Concorde service" (http://select.nytimes.com/gst/abstract.html?res=F10F15F6395416768FDDAE0994DA415B878BF1D3). *The New York Times,* 17 December 1977. 17 December 1977. . Retrieved 30 June 2011.

[105] "Concorde route cut" (http://news.google.co.uk/newspapers?id=eZckAAAAIBAJ&sjid=raQFAAAAIBAJ&pg=2330,3152191&dq=concorde+singapore+airlines&hl=en). *Montreal Gazette,* 16 September 1980. . Retrieved 30 June 2011.

[106] "French Concorde to Mexico City" (http://news.google.co.uk/newspapers?id=uekpAAAAIBAJ&sjid=4tIEAAAAIBAJ&pg=4845,5120670&dq=concorde+mexico&hl=en). *Daytona Beach Morning Journal,* 11 August 1978. . Retrieved 30 June 2011.

[107] "Supersonic Jet flights suspended" (http://news.google.co.uk/newspapers?id=TuwwAAAAIBAJ&sjid=8t8FAAAAIBAJ&pg=3054,6003139&dq=concorde+mexico&hl=en). *Daytona Beach Morning Journal,* 27 September 1982. . Retrieved 30 June 2011.

[108] "Air France offering 'New Year's Eve in Paris.'" (http://www.highbeam.com/doc/1G1-5212185.html). *PR Newswire,* 2 October 1987. . Retrieved 30 June 2011.

[109] Getze, John (10 February 1977). "Braniff seeks deal to fly Concorde in U.S" (http://pqasb.pqarchiver.com/latimes/access/646052352.html?dids=646052352:646052352&FMT=ABS&FMTS=ABS:AI&type=historic&date=Feb+10,+1977&author=&pub=Los+Angeles+Times&desc=Braniff+Seeks+Deal+to+Fly+Concorde+in+U.S.&pqatl=google). *Los Angeles Times,* 10 February 1977. .

[110] "Concorde flights to Texas Ok'd" (http://pqasb.pqarchiver.com/latimes/access/647634162.html?dids=647634162:647634162&FMT=CITE&FMTS=CITE:AI&type=historic&date=Jun+22,+1978&author=&pub=Los+Angeles+Times&desc=Concorde+Flights+to+Texas+OKd&pqatl=google). *Los Angeles Times,* 22 June 1978. 22 June 1978. . Retrieved 30 June 2011.

[111] "Concorde now reaping profits on N.Y. route" (http://news.google.co.uk/newspapers?id=VAQkAAAAIBAJ&sjid=MO4DAAAAIBAJ&pg=6783,4192544&dq=concorde+braniff&hl=en). *The Spokesman-Review,* 23 November 1979. . Retrieved 30 June 2011.

[112] "Braniff to halt US Concorde flights" (http://news.google.co.uk/newspapers?id=-2UaAAAAIBAJ&sjid=ACsEAAAAIBAJ&pg=6806,4059124&dq=concorde+braniff&hl=en). *Milwaukee Journal,* 16 April 1980. . Retrieved 30 June 2011.

[113] "Concorde flights between Texas and Europe end; Big Dreams at the start, $1,447 for flight to Paris" (http://select.nytimes.com/gst/abstract.html?res=F00911FA3C5C17728DDDA80894DE405B8084F1D3). *The New York Times,* 1 June 1980. 1 June 1980. . Retrieved 30 June 2011.

[114] *B.CAL appoints Concorde consultant.* (http://www.flightglobal.com/pdfarchive/view/1979/1979 - 0919.html) Flight International, 24 March 1979, p. 881.

[115] *Caledonian reports best results.* (http://www.flightglobal.com/pdfarchive/view/1979/1979 - 1617.html) Flight International, 12 May 1979, p. 1547.

[116] *New job for B.CAL Concorde man.* (http://www.flightglobal.com/pdfarchive/view/1979/1979 - 2664.html) Flight International, 21 July 1979, p. 158.

[117] *B.CAL to lease unsold Concorde?.* (http://www.flightglobal.com/pdfarchive/view/1979/1979 - 0541.html) Flight International, 24 February 1979, p. 517.

[118] *Tories support B.CAL's Concorde plan.* (http://www.flightglobal.com/pdfarchive/view/1979/1979 - 1182.html) Flight International, 14 April 1979, p. 1132.

[119] World News. (http://www.flightglobal.com/pdfarchive/view/1979/1979 - 1348.html) Flight International, 28 April 1979, p. 1286.

[120] *High Risk: The Politics of the Air,* Thomson, A., Sidgwick and Jackson, London, 1990, pp. 316–317.

[121] *B.CAL drops Concorde plans but asks for Hong Kong licence.* (http://www.flightglobal.com/pdfarchive/view/1979/1979 - 2433.html) Flight International, 30 June 1979, p. 2331.

[122] *BCAL's African Profit ....* (http://www.flightglobal.com/pdfarchive/view/1971/1971 - 2766.html) World News, Flight International, 23 December 1971, p. 994.

[123] *The Caledonian punchbag,* Flight International, 21 March 1987, p. 33.

[124] *Cheaper fares on Swedish routes?,* Air Transport, Flight International, 31 March 1979, p. 965 (http://www.flightglobal.com/pdfarchive/view/1979/1979 - 1007.html)

[125] *Two Concordes, A310s and 20 new routes in B.CAL package,* Air Transport, Flight International, 19 May 1979, p. 1637 (http://www.flightglobal.com/pdfarchive/view/1979/1979 - 1711.html)

[126] Backroom boys — Francis Spufford

[127] Greenberg, Peter (1 April 1984). "The plane fact is, Concorde has broken the profit barrier for the first time" (http://pqasb.pqarchiver.com/chicagotribune/access/632500842.html?dids=632500842:632500842&FMT=ABS&FMTS=ABS:AI&type=historic&date=Apr+01,+1984&author=&pub=Chicago+Tribune&desc=The+plane+fact+is,+Concorde+has+broken+the+profit+barrier+for+the+first+time&pqatl=google). *Chicago Tribune,* 1 April 1984. . Retrieved 30 June 2011.

[128] "The Concorde belies those who foresaw its extinction" (http://nl.newsbank.com/nl-search/we/Archives?p_product=PI&s_site=philly&p_multi=PI&p_theme=realcities&p_action=search&p_maxdocs=200&p_topdoc=1&p_text_direct-0=0EB29A63D671E93E&p_field_direct-0=document_id&p_perpage=10&p_sort=YMD_date:D&s_trackval=GooglePM). *Philadelphia Inquirer,* 26 January 1986. 26 January 1986. . Retrieved 30 June 2011.

[129] Arnold, James (10 October 2003). "Why economists don't fly Concorde" (http://news.bbc.co.uk/1/hi/business/2935337.stm). *BBC News,* 10 October 2003. .

[130] "Concorde to fly Miami-London route" (http://nl.newsbank.com/nl-search/we/Archives?p_product=MH&s_site=miami&p_multi=MH&p_theme=realcities&p_action=search&p_maxdocs=200&p_topdoc=1&p_text_direct-0=0EB35E9473D0D942&p_field_direct-0=document_id&p_perpage=10&p_sort=YMD_date:D&s_trackval=GooglePM). *Miami Herald,* 24 January 1984. 24 January 1984. . Retrieved 30 June 2011.

[131] "Cuts ground Concorde from Miami to London" (http://nl.newsbank.com/nl-search/we/Archives?p_product=MH&s_site=miami&p_multi=MH&p_theme=realcities&p_action=search&p_maxdocs=200&p_topdoc=1&p_text_direct-0=0EB342A3394157C7&p_field_direct-0=document_id&p_perpage=10&p_sort=YMD_date:D&s_trackval=GooglePM). *Miami Herald*. 13 January 1991. . Retrieved 30 June 2011.

[132] Jensen, Gregory. "After eight years, the Concorde flies to supersonic profit" (http://news.google.com/newspapers?id=R_oxAAAAIBAJ&sjid=NuQFAAAAIBAJ&pg=4247,5250844&dq=concorde+barbados&hl=en). *Reading Eagle*, 27 March 1984. .

[133] Greenberg, Peter S. (23 June 1985). "Flying via charter off to New Bustle" (http://pqasb.pqarchiver.com/latimes/access/64470984.html?dids=64470984:64470984&FMT=ABS&FMTS=ABS:FT&type=current&date=Jun+23,+1985&author=PETER+S.+GREENBERG&pub=Los+Angeles+Times+(pre-1997+Fulltext)&desc=Flying+Via+Charter+Off+to+New+Bustle&pqatl=google). *Los Angeles Times*, 23 June 1985. .

[134] Clark, Jay (18 January 1986). "Supersonic Concorde 10 years in service and still confounding critics" (http://pqasb.pqarchiver.com/thestar/access/474815901.html?dids=474815901:474815901&FMT=ABS&FMTS=ABS:FT&type=current&date=Jan+18,+1986&author=Jay+Clarke+Special+to+The+Star&pub=Toronto+Star&desc=Supersonic+Concorde+10+years+in+service+and+still+confounding+critics&pqatl=google). *Toronto Star*, 18 January 1986. .

[135] "Concorde money is in charters" (http://www.flightglobal.com/pdfarchive/view/1986/1986 - 0201.html). *Flight International*, 25 January 1986. . Retrieved 30 June 2011.

[136] Endres 2001, pp. 110–113.

[137] Cody, E. "French court holds Continental Airlines responsible for 2000 Concorde crash" (http://www.washingtonpost.com/wp-dyn/content/article/2010/12/06/AR2010120601740.html). *Washington Post*, 6 December 2010.

[138] Associated Press (30 November 2012). "French court overturns Concorde crash conviction" (http://www.nzherald.co.nz/world/news/article.cfm?c_id=2&objectid=10850947). New Zealand Herald. . Retrieved 30 November 2012.

[139] "Human Factor Issues Emerge from Concorde Crash Investigation" (http://findarticles.com/p/articles/mi_m0UBT/is_6_16/ai_82753359/). *Air Safety Week, 2002*. . Retrieved 1 February 2010.

[140] "Concorde's safety modifications" (http://news.bbc.co.uk/1/hi/uk/1442061.stm). *BBC News*, 17 July 2001. 17 July 2001. .

[141] "Concorde Completes Successful Test Flight" (http://www.foxnews.com/story/0,2933,29732,00.html). *Fox News*, 17 July 2001. 17 July 2001. . Retrieved 30 June 2011.

[142] "Concorde, 100 BA staff fly over Atlantic" (http://www.highbeam.com/doc/1G1-78118409.html). *United Press International*, 11 September 2001. .

[143] Williams, Timothy. "Concorde returns" (http://news.google.co.uk/newspapers?id=efkjAAAAIBAJ&sjid=yggEAAAAIBAJ&pg=3487,4067442&dq=concorde&hl=en). *Ocala Star-Banner*, 7 November 2001. .

[144] "Concorde 'back where she belongs'" (http://news.bbc.co.uk/1/hi/uk/1641466.stm). *BBC News*, 6 November 2001. 6 November 2001. .

[145] "Concorde grounded for good" (http://news.bbc.co.uk/2/hi/uk_news/2934257.stm). *BBC News*, 10 April 2003. 10 April 2003. . Retrieved 30 June 2011.

[146] Macdonald, Ian S. (1980). "New Aircraft: Where are we heading in the 1980s and 1990s" (http://www.emeraldinsight.com/Insight/viewContentItem.do;jsessionid=E7C9404F2C91DC6FD5BC8C45D43CED69?contentType=Article&hdAction=lnkpdf&contentId=1682944). *Aircraft Engineering and Aerospace Technology* (*MCB UP,*) **52** (7, 1980): 23–17. doi:10.1108/eb035646. ISSN 0002-2667. .

[147] Michaels, Danial (2 October 2003). "Final Boarding Call: As Concorde Departs, so do 3-Man Crews: In New Cockpits, Engineers are seen as Extra Baggage" (http://online.wsj.com/article/SB106504293992222300.html?mod=googlewsj). *Wall Street Journal*, 2 October 2003. .

[148] Montague, Simon (6 May 2003). "Branson's Concorde bid rejected" (http://news.bbc.co.uk/1/hi/uk_politics/3005705.stm). *BBC News*, 6 May 2003. .

[149] "Branson accuses 'sad' Government of washing its hands of Concorde" (http://www.highbeam.com/doc/1G1-109201077.html). *Western Mail*, 24 October 2003. . Retrieved 30 June 2011.

[150] "Concorde not to fly at air shows" (http://edition.cnn.com/2003/WORLD/europe/10/30/concorde.nofly.reut/). *CNN*, 30 October 2003. 30 October 2003. . Retrieved 30 June 2011.

[151] Simpkins, Edward (15 June 2003). "Buffett vehicle to follow in Concorde's slipstream" (http://www.telegraph.co.uk/finance/2855044/Buffett-vehicle-to-follow-in-Concordes-slipstream.html). London: *The Daily Telegraph*, 15 June 2003. .

[152] Quote: "Airbus, the manufacturer of Concorde, has said it is becoming uneconomic to maintain the ageing craft and that it will no longer provide spare parts for it."

[153] "Concorde: An Untimely and Unnecessary Demise" (http://www.thetravelinsider.info/2003/0411.htm). *Travel Insider*, 2003. . Retrieved 13 January 2010.

[154] Lewis, Rob. *Supersonic Secrets: The Unofficial Biography of the Concorde*. London: Expose, a division of Secret Books Limited, 2003. ISBN 0-9546617-0-2.

[155] Pevsner, Donald L.. "The Betrayal of Concorde" (http://www.concorde-spirit-tours.com/concorde.htm). *Concorde Spirit Tours*. . Retrieved 13 January 2010.

[156] "I beat 20m to land last pounds 10 seats on Concorde; Joe grabs bargain of a lifetime." (http://www.thefreelibrary.com/I+BEAT+20m+TO+LAND+LAST+pounds+10+SEATS+ON+CONCORDE;+Joe+grabs...-a061046466) *Scottish Daily Record & Sunday (Glasgow,*

Scotland), 13 February 1997. Retrieved: 13 November 2012.

[157] Lemel, Laurent. "Concorde makes Final Flight from Paris to New York". *Associated Press,* 30 May 2003.

[158] "French Concorde bids adieu" (http://news.bbc.co.uk/1/hi/world/europe/2952594.stm). *BBC News,* 31 May 2003. 31 May 2003. .

[159] "Jetting off" (http://www.highbeam.com/doc/1G1-104484558.html). *Daily Mirror,* 28 June 2003. . Retrieved 30 June 2011.

[160] "3,500 due at UK Concorde auction" (http://news.bbc.co.uk/1/hi/uk/3246326.stm). *BBC News,* 30 November 2003. 30 November 2003. .

[161] "Miscellaneous brief articles – Business & Industry" (http://www.alacrastore.com/storecontent/Business-and-Industry/105669362). *Flight International.* 15 July 2003. . Retrieved 30 June 2011.

[162] "Pourquoi n'a-t-on pas sauvé le Concorde?" (http://www.tourmag.com/Pourquoi-n-a-t-on-pas-sauve-le-Concorde_a7.html). *TourMag.com,* 24 March 2010. . Retrieved 30 June 2011.

[163] Webster, Ben. "This is not a flight of fancy: Volunteers say Concorde can realise an Olympic dream if BA will help". *The Times,* 31 May 2006.

[164] "Museum Sinsheim" (http://sinsheim.technik-museum.de/concorde-f-bvfb/). *Museum Sinsheim.* . Retrieved 26 June 2010.

[165] Atchison, Marc (2 October 2003). "Concorde's supersonic swan song; Star writer aboard for jet's farewell trip to Toronto Transatlantic sound-breaker a vision of grace" (http://pqasb.pqarchiver.com/thestar/access/417072421.html?dids=417072421:417072421&FMT=ABS&FMTS=ABS:FT&type=current&date=Oct+02,+2003&author=Marc+Atchison&pub=Toronto+Star&desc=Concorde's+supersonic+swan+song;+Star+writer+aboard+for+jet's+farewell+trip+to+Toronto+Transatlantic+sound-breaker+a+vision+of+grace&pqatl=google). *Toronto Star,* 2 October 2003. .

[166] Ramstack, Tom. "Final flight: British Airways Concorde lands locally for last time" (http://goliath.ecnext.com/coms2/gi_0199-3250817/Final-flight-British-Airways-Concorde.html). *Washington Post,* 15 October 2003. .

[167] "Concorde establishes London-to-U.S. record" (http://www.accessmylibrary.com/coms2/summary_0286-24654137_ITM). *America's Intelligence Wire,* 9 October 2003. 9 October 2003. . Retrieved 30 June 2011.

[168] "SST makes record flight" (http://nl.newsbank.com/nl-search/we/Archives?p_product=SL&p_theme=sl&p_action=search&p_maxdocs=200&p_topdoc=1&p_text_direct-0=0EB04F816AA79462&p_field_direct-0=document_id&p_perpage=10&p_sort=YMD_date:D&s_trackval=GooglePM). *St Louis Post,* 9 February 1996. 9 February 1996. . Retrieved 30 June 2011.

[169] "Exhibits: Concorde". (http://www.intrepidmuseum.org/The-Intrepid-Experience/Exhibits/Concorde.aspx) *Intrepidmuseum.org.* Retrieved 6 March 2011.

[170] "Last Concorde lands" (http://news.bbc.co.uk/2/hi/uk_news/england/bristol/somerset/3238674.stm). *BBC News,*. 27 November 2003. . Retrieved 30 June 2011.

[171] "Concorde enjoys Cardiff farewell" (http://news.bbc.co.uk/1/hi/wales/3205915.stm). *BBC News.* 23 October 2003. 23 October 2003. .

[172] "Concorde – The Farewell – A collection of the final flights of the last days of Concorde" (http://www.amazon.co.uk/dp/B0001DI57G). *Simply Media,* 19 April 2004. . Retrieved 30 June 2011.

[173] Magoolaghan, Brian. "The Concorde Makes A Comeback" (http://www.rockawave.com/news/2003-10-31/Front_Page/028.html). *Wave of Long Island,* 31 October 2003. . Retrieved 30 June 2011.

[174] Lawless, Jill. "Final Concorde flight lands at Heathrow". *Associated Press,* 26 October 2003.

[175] "End of an era for Concorde" (http://news.bbc.co.uk/1/hi/uk/3211053.stm). *BBC News,*. 24 October 2003. .

[176] Thousands welcome Concorde to Seattle (http://www.seattlepi.com/local/article/Thousands-welcome-Concorde-to-Seattle-1128999.php)

[177] Robert Bogash (http://rbogash.com/Concorde.html)

[178] "BBC News: Last Concorde lands" (http://news.bbc.co.uk/1/hi/england/bristol/3238674.stm). *BBC News,*. 27 November 2003. .

[179] "Will Concorde ever come out of retirement – e.g. for a Coronation flypast or airshows?" (http://www.britishairways.com/concorde/faq.html#17). *British Airways.* . Retrieved 14 January 2010.

[180] "Concorde nose cone sells for half-million at auction" (http://www.usatoday.com/travel/news/2003-12-02-concorde_x.htm). *USA Today,* 12 December 2003. 2 December 2003. . Retrieved 30 June 2011.

[181] "Concorde Memorabilia Auction" (http://www.highbeam.com/doc/1P1-87889450.html). *Getty Images,* 1 December 2003. . Retrieved 30 June 2011.

[182] "Heathrow Concorde model removed" (http://news.bbc.co.uk/1/hi/england/london/6509667.stm). *BBC News,*. 30 March 2007. .

[183] "The Brooklands Concorde Project" (http://www.concordeproject.com/). *Brooklands Museum.* . Retrieved 15 January 2010.

[184] "Air France Concorde to taxi again under own power" (http://www.flightglobal.com/articles/2010/02/05/338114/air-france-concorde-to-taxi-again-under-own-power.html). *Flightglobal,* 5 February 2010. . Retrieved 5 February 2010.

[185] "Work starts in £15m plan to get Concorde Flying" (http://news.bbc.co.uk/1/hi/world/europe/8712806.stm). *BBC News,*29 May 2010. 29 May 2010. . Retrieved 29 May 2010.

[186] "Soviet Union: Christening the Concordski" (http://www.time.com/time/magazine/article/0,9171,912014,00.html). *Time,* 14 November 1977. 14 November 1977. . Retrieved 30 June 2011.

[187] Gordon, Yefim. *Tupolev Tu-144.* London: Midland, 2006. ISBN 1-85780-216-0.

[188] "'Concordski' designer dies" (http://news.bbc.co.uk/1/hi/world/europe/1328132.stm). *BBC News,* 13 May 2001. 13 May 2001. .

[189] Hess, John L (26 May 1971). "Soviet SST, in Its First Flight to the West, Arrives in Paris for Air Show" (http://select.nytimes.com/gst/abstract.html?res=F4091EFB3954127B93C4AB178ED85F458785F9). *The New York Times,* 26 May 1971. . Retrieved 30 June 2011.

[190] Deruaz, George. "Soviet SST stalls, dives into towns" (http://news.google.co.uk/newspapers?id=2JwMAAAAIBAJ&sjid=5l8DAAAAIBAJ&pg=4974,3987586&dq=tu+144&hl=en). *St. Petersburg Times,* 4 June 1973. . Retrieved 30 June 2011.

[191] "Pride of Soviet air fleet explodes during exhibition" (http://news.google.co.uk/newspapers?id=DBsfAAAAIBAJ&sjid=Ko0EAAAAIBAJ&pg=3185,500285&dq=tu+144&hl=en). *Sarasota Journal,* 4 June 1973. . Retrieved 30 June 2011.

[192] Fisher, Dan (27 October 1978). "Russia confirms crash of Supersonic Airliner in test: Latest failure of Trouble-plagued TU-144 seen as blow to Soviet hopes of expanding industry" (http://pqasb.pqarchiver.com/latimes/access/659730132.html?dids=659730132:659730132&FMT=ABS&FMTS=ABS:AI&type=historic&date=Oct+27,+1978&author=&pub=Los+Angeles+Times&desc=Russia+Confirms+Crash+of+Supersonic+Airliner+in+Test&pqatl=google). *Los Angeles Times,* 27 October 1978. .

[193] "ASN Aircraft accident Tupolev 144D CCCP-77111 Yegoryevsk" (http://aviation-safety.net/database/record.php?id=19780523-1). Flight Safety Foundation. . Retrieved 1 July 2011.

[194] "The Tu-144: the future that never was" (http://en.rian.ru/analysis/20080603/109170103.html). *RIA Novosti,* 3 January 2008. . Retrieved 30 June 2011.

[195] Fridlyander, Iosif. "Sad Epic of the Tu-144." *Messenger of Russian Academy of Sciences*, №1, 2002 (in Russian: И.Н. Фридляндер, "Печальная эпопея Ту-144" (http://vivovoco.rsl.ru/VV/JOURNAL/VRAN/02_01/FRID.HTM), Вестник РАН, №1, 2002.

[196] "The United States SST Contenders" (http://www.flightglobal.com/pdfarchive/view/1964/1964 - 0377.html). *Flight International,* 13 February 1964. pp. 234–235. . Retrieved 30 June 2011.

[197] Winchester 2005, p. 84.

[198] Lyons, Richard D (5 January 1969). "The Russians Lead With the SST..." (http://select.nytimes.com/gst/abstract.html?res=F10D11F6395F127A93C7A9178AD85F4D8685F9). *The New York Times,* 5 January 1969. . Retrieved 30 June 2011.

[199] "B-58's Sonic Boom Rattles Kentuckians" (http://pqasb.pqarchiver.com/chicagotribune/access/574189232.html?dids=574189232:574189232&FMT=CITE&FMTS=CITE:AI&date=Dec+19,+1961&author=&pub=Chicago+Daily+Tribune&desc=B-58's+SONIC+BOOM+RATTLES+KENTUCKIANS&pqatl=google). *Chicago Daily Tribune,* 19 December 1961. 19 December 1961. . Retrieved 30 June 2011.

[200] "The Nation: Showdown on the SST" (http://www.time.com/time/magazine/article/0,9171,944291,00.html). *Time,* 29 May 1971. 29 March 1971. . Retrieved 30 June 2011.

[201] "Tu-160 Blackjack Strategic Bomber, Russia" (http://www.airforce-technology.com/projects/tu160/). *airforce-technology.com.* . Retrieved 16 May 2010.

[202] "Here Comes the Concorde, Maybe" (http://www.time.com/time/magazine/article/0,9171,918012,00.html). *Time,* 16 February 1976. 16 February 1976. . Retrieved 30 June 2011.

[203] Allen, Robert M.. *Legal and Environmental ramifications of the Concorde* (http://heinonline.org/HOL/LandingPage?collection=journals&handle=hein.journals/jalc42&div=28&id=&page=). J. Air L. & Com., 1976. . Retrieved 30 June 2011.

[204] Hock, R.; R. Hawkins. *Recent studies into Concorde noise reduction* (http://md1.csa.com/partners/viewrecord.php?requester=gs&collection=TRD&recid=N7422661AH&q=Concorde+noise&uid=788872723&setcookie=yes). *AGARD Noise Mech,* 1974. p. 14. . Retrieved 30 June 2011.

[205] Muss, Joshua A.. *Aircraft Noise: Federal pre-emption of Local Control, Concorde and other recent cases* (http://heinonline.org/HOL/LandingPage?collection=journals&handle=hein.journals/jalc43&div=45&id=&page=). *J. Air L. & Com,* 1977. . Retrieved 30 June 2011.

[206] Endres 2001, p 90.

[207] "Reducing noise with type 28 nozzle" (http://www.emeraldinsight.com/Insight/viewContentItem.do?contentType=Article&hdAction=lnkpdf&contentId=1682311). *Aircraft Engineering and Aerospace Technology* (*MCB UP*) **45** (4, 1973): 14. 1973. doi:10.1108/eb035013. .

[208] Fahey, D. W.; et al. (1995). "Emission Measurements of the Concorde Supersonic Aircraft in the Lower Stratosphere" (http://www.sciencemag.org/cgi/content/abstract/sci;270/5233/70). *Science* **270** (5233): 70. Bibcode 1995Sci...270...70F. doi:10.1126/science.270.5233.70. .

[209] "Increase in supersonic jets could be threat to ozone U-2 plane trails Concorde, studies exhaust particles." (http://articles.baltimoresun.com/1995-10-08/news/1995281022_1_ozone-sulfur-exhaust-particles) *Baltimore Sun,* October 8, 1995.

[210] Anderson, Jon. *Decision Analysis in Environmental Decisionmaking: Improving the Concorde Balance* (http://heinonline.org/HOL/LandingPage?collection=journals&handle=hein.journals/cjel5&div=9&id=&page=). *HeinOnline,* 1978. . Retrieved 30 June 2011.

[211] "Train à grande vitesse causes distress" (http://www.esemag.com/1101/1101ed.html). *Environmental Science and Engineering,* November 2001. .

[212] "National and regional tranquillity maps" (http://www.cpre.org.uk/campaigns/landscape/tranquillity/national-and-regional-tranquillity-maps). *Campaign to Protect Rural England.* . Retrieved 25 April 2010.

[213] "British Airways Concorde is expected to begin flying passengers again in next 6 weeks" (http://www.alacrastore.com/storecontent/Business-and-Industry/24962872). *Dallas Morning News,* 23 August 2001. . Retrieved 30 June 2011.

[214] "Farewell to Concorde" (http://news.bbc.co.uk/1/hi/in_depth/uk/2003/concorde_retirement/default.stm). *BBC News,* 15 August 2007. 15 August 2007. . Retrieved 15 January 2010.

[215] Oxford Language Dictionaries Online – French Resources (http://www.oxfordlanguagedictionaries.com/Public/PublicResources.html?direction=b-fr-en&sp=S/oldo/resources/fr/French-supp-05.xml): Glossary of Grammatical Terms

[216] Centre National de Ressources Textuelles et Lexicales (http://www.cnrtl.fr/lexicographie/academie9/le) – Définition de LE, LA: article défini, II.3

[217] Reverso Dictionnaire: La majuscule dans les noms propres (http://grammaire.reverso.net/5_2_01_la_majuscule_dans_les_noms_propres.shtml) ("Capital letters within proper names")

[218] **concorde** *s.f.* concord, unity, harmony, peace.Ferrar 1980, p. 114.

[219] "Celebrate Concorde: videos" (http://www.britishairways.com/travel/concvidhome/public/en_gb). *British Airways*. . Retrieved 15 January 2010.

[220] "Documentary of British Airways Concorde introduction" (http://www.youtube.com/watch?v=Ro3QVMCG-aY). *youtube.com*. .

[221] Quote: Raymond Baxter commentating as Concorde flies for first time: "She rolls ... She flies!"

[222] "Red Arrows fly into Scotland" (http://www.highbeam.com/doc/1G1-62676656.html). *Daily Record,* 12 June 2000. . Retrieved 30 June 2011.

[223] "Million turn out to crown Queen's Jubilee" (http://www.telegraph.co.uk/education/3297816/Million-turn-out-to-crown-Queens-Jubilee.html). London: *Daily Telegraph,* 4 June 2002. 4 June 2002. . Retrieved 1 April 2010.

[224] Laville, Sandra (24 October 2003). "Chaos fear at Concorde farewell" (http://www.telegraph.co.uk/news/uknews/1444939/Chaos-fear-at-Concorde-farewell.html). London: *Daily Telegraph,* 24 October 2003. .

[225] "Air France Concorde sets round-the-world speed record" (http://www.allbusiness.com/transportation/air-transportation-airports/7142582-1.html). *Business Wire,* 16 August 1995. . Retrieved 30 June 2011.

[226] "French Concorde to attempt round-the-world record". *Anchorage Daily News,* 12 October 1992.

[227] "Concorde jets occupants on record ride". *Deseret News,* 17 August 1995.

[228] "Rolls-Royce SNECMA Olympus". *Janes,* 25 July 2000.

[229] "Museum of Flight" (http://www.museumofflight.org/). Museum of Flight. . Retrieved 21 August 2011.

[230] "French transport chief speculates about new-generation Concorde" (http://nl.newsbank.com/nl-search/we/Archives?p_product=NewsLibrary&p_multi=APAB&d_place=APAB&p_theme=newslibrary2&p_action=search&p_maxdocs=200&p_topdoc=1&p_text_direct-0=0F89C550A53D43FD&p_field_direct-0=document_id&p_perpage=10&p_sort=YMD_date:D&s_trackval=GooglePM). *Associated Press,* 17 August 2000. 17 August 2000. . Retrieved 30 June 2011.

[231] Cody, Edward (10 May 1990). "Partnership gears up for Concorde sequel; British, French firms sign plane pact" (http://pqasb.pqarchiver.com/washingtonpost/access/72587838.html?dids=72587838:72587838&FMT=ABS&FMTS=ABS:FT&type=current&date=May+10,+1990&author=Edward+Cody&pub=The+Washington+Post+(pre-1997+Fulltext)&desc=Partnership+Gears+Up+For+Concorde+Sequel;+British,+French+Firms+Sign+Plane+Pact&pqatl=google). *Washington Post,* 10 May 1990. .

[232] "Firm considers 'son of Concorde'" (http://news.bbc.co.uk/1/hi/business/3231354.stm). *BBC News,* 23 November 2003. 23 November 2003. .

[233] "Japan, France working on new supersonic jet" (http://www.msnbc.msn.com/id/8229113/). *MSNBC,* 15 June 2005. . Retrieved 30 June 2011.

[234] "Japan tests supersonic jet model" (http://news.bbc.co.uk/2/hi/science/nature/4325634.stm). *BBC News,* 10 October 2005. 10 October 2005. . Retrieved 30 June 2011.

[235] Kaminski-Morrow, David (19 June 2011). "Paris: EADS details near-hypersonic transport concept" (http://www.flightglobal.com/articles/2011/06/19/358271/paris-eads-details-near-hypersonic-transport-concept.html). Flight Daily News. .

[236] Lichfield, John (20 June 2011). "London to New York in 90 minutes: is this the Concorde of the future?" (http://www.independent.co.uk/news/science/london-to-new-york-in-90-minutes-is-this-the-concorde-of-the-future-2299925.html). *The Independent*. .

[237] "LAPCAT aims at supersonic civil aviation" (http://www.gizmowatch.com/entry/lapcat-aims-at-supersonic-civil-aviation/). *Gizmo Watch,* 30 August 2007. . Retrieved 3 July 2009.

[238] "LAPCAT II – Facts and figures" (http://www.esa.int/esaMI/Space_Engineering/SEMD0G5KXMF_0.html). *European Space Agency,* 17 November 2009. . Retrieved 10 August 2010.

[239] O'Connell, Dominic (18 May 2008). "Orders for Aerion's Concorde executive jet are more than $3 billion" (http://business.timesonline.co.uk/tol/business/industry_sectors/transport/article3953823.ece). London: *The Times,* 18 May 2008. .

[240] Nanveen. "More details emerge on the $80 million Aerion Supersonic Business Jet." (http://www.myduckeggs.com/category/supersonic-aerospace-international/) *myduckeggs.com*, 26 July 2010. Retrieved 28 July 2010.

[241] Hagerman, Eric. "Supersonic jet promises to fly nearly silent" (http://www.cnn.com/2007/TECH/02/16/quiet.supersonic/index.html). *CNN,* 16 February 2007. .

[242] Kelly 2005, p. 52.

[243] "Comparison with Tu-144." (http://www.richard-seaman.com/Aircraft/Museums/Monino/Highlights/index.html) *Richard Seaman aircraft museum*. Retrieved 15 January 2010.

[244] BA and AF Concordes originally had 100 seats, with AF removing 8 seats after the safety modifications of CY2000–2001 because of weight considerations.

Citations

Bibliography

- Beniada, Frederic. *Concorde*. Minneapolis, Minnesota: Zenith Press, 2006. ISBN 0-7603-2703-3.
- Calvert, Brian. *Flying Concorde: The Full Story*. London: Crowood Press, 2002. ISBN 1-84037-352-0.
- Deregel, Xavier; Jean-Philippe Lemaire. *Concorde Passion*. New York: LBM, 2009. ISBN 2-915347-73-5.

- Endres, Günter. *Concorde*. St Paul, Minnesota: MBI Publishing Company, 2001. ISBN 0-7603-1195-1.
- Ferrar, Henry (ed.). *The Concise Oxford French-English Dictionary*. New York: Oxford University Press, 1980. ISBN 0-19-864157-5.
- Frawley, Gerald (2003). *The International Directory of Civil Aircraft, 2003/2004*. Aerospace Publications. ISBN 1-875671-58-7.
- Gunn, John (2010). *Crowded Skies*. Turnkey Productions. ISBN 978-0-646-54973-6.
- Kelly, Neil. *The Concorde Story: 34 Years of Supersonic Air Travel*. Surrey, UK: Merchant Book Company Ltd, 2005. ISBN 1-904779-05-0.
- Knight, Geoffrey. *Concorde: The Inside Story*. London: Weidenfeld and Nicolson, 1976. ISBN 0-297-77114-0.
- McIntyre, Ian. *Dogfight: The Transatlantic Battle over Airbus*. Westport, Connecticut: Praeger Publishers, 1992. ISBN 0-275-94278-3.
- Nunn, John Francis. *Nunn's Applied Respiratory Physiology*. Burlington, Maryland: Butterworth-Heineman, 1993. ISBN 0-7506-1336-X.
- Owen, Kenneth (2001). Concorde: Story of a Supersonic Pioneer (http://books.google.com/?id=VzNUJlX7CXoC&printsec=frontcover&dq=Concorde:+story+of+a+supersonic+pioneer#v=onepage&q&f=false). London: Science Museum, 2001. ISBN 978-1-900747-42-4.
- Orlebar, Christopher. *The Concorde Story*. Oxford, UK: Osprey Publishing, 2004. ISBN 1-85532-667-1.
- Ross, Douglas (March 1978). *The Concorde Compromise: The Politics of Decision-making*. Bulletin of the Atomic Scientists. pp. 46–53.
- Schrader, Richard K. *Concorde: The Full Story of the Anglo-French SST*. Kent, UK: Pictorial Histories Pub. Co., 1989. ISBN 0-929521-16-1.
- Taylor, John W. R. (1965). *Jane's All The World's Aircraft 1965–66*. Marston.
- Towey, Barrie (ed.). *Jet Airliners of the World 1949–2007*. Tunbridge Wells, Kent, UK: Air-Britain (Historians) Ltd, 2007. ISBN 0-85130-348-X.
- Winchester, Jim. *The World's Worst Aircraft: From Pioneering Failures to Multimillion Dollar Disasters*. London: Amber Books Ltd, 2005. ISBN 1-904687-34-2.
- Winchester, Jim (2005). *X-Planes and Prototypes: From Nazi Secret Weapons to the Warplanes of the Future*. Amber Books Ltd. ISBN 1-84013-815-7.

External links

- British Airways Concorde page (http://www.britishairways.com/concorde/index.html)
- Braniff Airways Concorde page (http://www.braniffpages.com/concorde.html)
- Design Museum (UK) Concorde page (http://www.designmuseum.org/design/concorde)
- Concorde Experience at the National Museum of Flight, Scotland (http://www.nms.ac.uk/our_museums/museum_of_flight/concorde.aspx)
- Concorde F-BVFB at the Auto & Technik Museum Sinsheim (http://sinsheim.technik-museum.de/en/concorde-f-bvfb)
- Supersonic: Tribute to Concorde (http://www.life.com/image/first/in-gallery/24961/supersonic-tribute-to-concorde) – a slideshow by *Life magazine*
- Pathé film of 002's maiden flight from Filton to RAF Fairford on 9 April 1967 (http://www.britishpathe.com/record.php?id=45713)
- First Supersonic Transatlantic Commercial Flight Lands at Dulles (1976) (http://ghostsofdc.org/2012/01/30/first-supersonic-transatlantic-commercial-flight-lands-at-dulles-1976/)

BT_Group

Type	Public limited company
Traded as	LSE: BT.A [1] NYSE: BT [2]
Industry	Telecommunications
Predecessor(s)	Electric Telegraph Company Post Office Telecommunications
Founded	1846 (City of London)
Founder(s)	William Cooke John Ricardo
Headquarters	**BT Centre**, London, United Kingdom
Area served	Worldwide
Key people	Sir Michael Rake (Chairman) Ian Livingston (Chief Executive)
Products	Fixedline telephony Mobile telephony Broadband internet Digital television IT services
Revenue	£18.897 billion (2011/12)[3]
Operating income	£2.919 billion (2011/12)[3]
Profit	£2.003 billion (2011/12)[3]
Employees	89,000 (at 31 March 2012)[3]
Website	btplc.com [4]

BT Group plc (trading as **BT**) is a British multinational telecommunications services company headquartered in London, United Kingdom. It is one of the largest telecommunications services companies in the world and has operations in over 170 countries.[5] Through its BT Global Services division it is a major supplier of telecoms services to corporate and government customers worldwide.[6] Its BT Retail division is a leading supplier of telephony, broadband and subscription television services in the UK, with over 18 million customers.[7] During the summer of 2012, BT was an official London 2012 Olympic Partner where its BT Global Services division constructed and operated the 2012 Summer Olympics' data network across 94 locations, the first Olympic/Paralympic Games to completely rely upon VoIP for voice communications as well as the first to offer WiFi to the public.[8] [9]

BT's origins date back to the founding of the Electric Telegraph Company in 1846, the first company in the world to develop a nationwide communications network. In 1912, the General Post Office, a government department, became the monopoly telecoms supplier in the UK. The Post Office Act of 1969 led to the GPO becoming a public corporation. **British Telecom** was formed in 1980, and became independent of the Post Office in 1981. British Telecom was privatised in 1984, with more than 50 per cent of its shares sold to investors. The Government sold its remaining stake in further share sales in 1991 and 1993.

BT has a primary listing on the London Stock Exchange and is a constituent of the FTSE 100 Index. It had a market capitalisation of approximately £15 billion as of 23 December 2011, the 27th-largest of any company with a primary listing on the London Stock Exchange.[10] It has a secondary listing on the New York Stock Exchange.

History

BT is the world's oldest telecommunications company. The company's origins date back to the establishment of the first telecommunications companies in the United Kingdom. Among them was the first commercial telegraph service, the Electric Telegraph Company, established in 1846. As these companies amalgamated and were taken over or collapsed, the remaining companies were transferred to state control under the Post Office. These companies were merged and rebranded as British Telecommunications.[11]

1878 to 1969

From 1878, the telephone service in the United Kingdom was provided by private sector companies such as the National Telephone Company, and later by the General Post Office. In 1896, the National Telephone Company was taken over by the General Post Office. In 1912, it became the primary supplier of telecommunications services, after the Post Office took over the private sector telephone service in the UK, except for a few local authority services. Those services all folded within a few years, the sole exception being Kingston upon Hull, where the telephone department became present day KCOM Group.[11] Telegraph and Telephone services became the exclusive responsibility of the Post Office Engineering department.

The BT Tower, constructed between 1961 and 1964.

Converting the Post Office into a nationalised industry, as opposed to a governmental department, was first discussed in 1932 by Lord Wolmer. His ideas were recorded in the book *Post Office Reform*. In 1932, the Bridgeman Committee was formed, "to enquire and report as to whether any changes to the constitution, status or system of organisation of the Post Office would be in the public interest". The Committee's report was rejected. In 1961, the subject received further attention, the proposals were ignored. The Post Office remained a department of central government, with the Postmaster General sitting in Cabinet as a Secretary of State.[11]

In March 1965, Tony Benn, the acting Postmaster General, wrote to the Prime Minister, Harold Wilson, proposing that studies be undertaken aimed at converting the Post Office into a nationalised industry. A party was established to look into the advantages and disadvantages of the proposal, and its findings were found to be favourable enough for the Government to re-establish a Steering Group on the Organisation of the Post Office. After some initial deliberations that the business should be divided into five divisions; Post, Telecommunications, Savings, Giro and National Data Processing Services, it was decided that there should be two: *Post* and *Telecommunications*. These events finally resulted in the introduction of the Post Office Act, 1969.[11]

On 1 October 1969, under the Post Office Act, 1969, the Post Office ceased to be a government department and it became established as a public corporation. The Act gave the Post Office the exclusive privilege of operating telecommunications systems with listed powers to authorise others to run such systems. Effectively, the General Post Office retained its telecommunications monopoly.[11]

1969 to 1982

In 1977, the Carter Committee Report recommended a further division of the two main services and for their relocation under two individual corporations. The findings contained in the report led to the renaming of Post Office Telecommunications as British Telecom in 1980, although it remained part of the Post Office.

The British Telecommunications Act, 1981 transferred the responsibility for telecommunications services from the Post Office, creating two separate corporations, Post Office Ltd. and British Telecom. At this time the first steps were taken to introduce competition into the UK telecommunications industry. In particular, the Act empowered the Secretary of State for Trade and Industry, as well as British Telecom, to license other operators to run public telecommunications systems. Additionally, a framework was established which enabled the Secretary of State to set standards with the British Standards Institution (BSI) for apparatus supplied to the public by third parties, and had the effect of requiring British Telecom to connect approved apparatus to its systems.

The Secretary of State made use of these new powers and began the process of opening up the apparatus supply market, where a phased programme of liberalisation was started in 1981. In 1982, a licence was granted to Cable & Wireless to run a public telecommunications network through its subsidiary, Mercury Communications Ltd.

1982 to 1991

On 19 July 1982, the Government formally announced its intention to privatise British Telecom with the sale of up to 51 per cent of the company's shares to private investors. This intention was confirmed by the passing of the Telecommunications Act, 1984, which received Royal Assent on 12 April that year. The transfer to British Telecommunications plc of the business of British Telecom, the statutory corporation, took place on 6 August 1984 and, on 20 November 1984, more than 50 per cent of British Telecom shares were sold to the public. At the time, this was the largest share issue in the world.[12]

The new legislation had to enable British Telecom to become more responsive to competition in the UK and to expand its operations globally. Commercial freedom granted to British Telecom allowed it to enter into new joint ventures and, if it so decided, to engage in the manufacture of its own apparatus.

The company's transfer into the private sector continued in December 1991 when the Government sold around half its remaining holding of 47.6 per cent of shares, reducing its stake to 21.8 per cent. Substantially all the government's remaining shares were sold in a third flotation in July 1993, raising £5 billion for the Treasury and introducing 750,000 new shareholders to the company.

The 1984 Act also abolished British Telecom's exclusive privilege of running telecommunications systems and established a framework to safeguard the workings of competition. This meant that British Telecom finally lost its monopoly in running telecommunications systems, which it had technically retained under the 1981 Act despite the Secretary of State's licensing powers. It now required a licence in the same way as any other telecommunications operator. The principal licence granted to British Telecom laid down strict and extensive conditions affecting the range of its activities, including those of manufacture and supply of apparatus.

The next major development for British Telecom, and a move towards a more open market in telecommunications, occurred in 1991. On 5 March, the Government's White Paper, Competition and Choice: Telecommunications Policy for the 1990s, was issued. In effect, it ended the duopoly which had been shared by British Telecom and Mercury Communications in the UK since November 1983 and the build up to privatisation. The new policy enabled customers to acquire telecommunications services from competing providers using a variety of technologies. Independent 'retail' companies were permitted to bulk-buy telecommunications capacity and sell it in packages to business and domestic users. The White Paper was endorsed by British Telecom, the new policy enabling the company to compete freely and more effectively by offering flexible pricing packages to meet the needs of different types of customer.

The BT Centre, completed in 1985.

1991 to 2001

On 2 April 1991, the company unveiled a new trading name, BT, a new corporate identity and a new organisational structure. This structure focused on specific market sectors, reflecting the needs of different customers – the individual, the small business or the multinational corporation. The reorganisation was named Project Sovereign to reflect the company's commitment to meetings customers' needs – 'the customer is King'. Together with a succession of strategic alliances with telecommunications companies worldwide, these changes gave BT the means to expand into overseas markets.

In December 2000, following modifications to BT's licence in April 2000, BT offered local loop unbundling (LLU) to other telecommunications operators, enabling them to use BT's copper local loops (the connection between the customer's premises and the exchange) to connect directly with their customers. By the end of August 2005, 105,055 lines had been unbundled.

Concert

In June 1994, BT and MCI Communication Corporation, the second largest carrier of long distance telecommunications services in the US, launched Concert Communications Services, a $1 billion joint venture company. This alliance gave BT and MCI a global network for providing end-to-end connectivity for advanced business services. Concert was the first company to provide a single source, broad portfolio of global communications services for multinational customers. On 3 November 1996, BT and MCI announced they had entered into a merger agreement to create a global telecommunications company called Concert plc, to be incorporated in the UK, with headquarters in both London and Washington DC. As part of the alliance BT acquired a 20 per cent holding in MCI. Nevertheless, following US carrier WorldCom's rival bid for MCI on 1 October 1997, BT ultimately decided in November, to sell its stake in MCI to WorldCom for $7 billion. The deal with WorldCom resulted in a profit of more than $2 billion on BT's original investment in MCI, with an additional $465 million severance fee for the break-up of the proposed merger.

In it was announced that BT and AT&T were to unwind Concert, returning its businesses, customer accounts and networks to the two parent companies (this was completed in April 2002).

O2

In 1985, Cellnet was launched as a subsidiary of Telecom Securicor Cellular Radio Limited, a 60:40 venture between British Telecommunications and Securicor respectively.

In 1999, BT purchased Securicor's shares in Cellnet for £3.15 billion, Securicor originally invested £4 million in Cellnet in 1983, the company was later rebranded as BT Cellnet, and it became a part of BT Wireless, a group of subsidiary companies owned by BT.[13]

In October 2001, at a general meeting held in Birmingham, 4.297 billion British Telecommunications shares voted in favour of the demerger, and 0.67 million voted against.[14] In 2001, BT Cellnet demerged from BT and was relaunched on 1 May 2002 as O2.

In February 2004, the company was subject to rumours of take-over bids from KPN, the Dutch telecommunications group though these never came to fruition. In 2005 rumours of KPN (and several other international network operators) bidding for the various branches of O2 persisted, though following the company's posting of better than expected results so soon after demerging some speculation shifted to O2 itself making bids for other businesses.

In March 2005, the company underwent a corporate reorganisation, that saw mmO2 plc being de-listed from the London Stock Exchange and acquired (via a share swap) by a new company, O2 plc, which was listed on the London Stock Exchange in its place. The summer of 2005 saw the company under threat of crippling strike action by Communication Workers Union members following disputes over pay and remuneration, though unrelated to earlier job losses as part of the reorganisation. This was resolved without strikes taking place after several months of negotiations between the company and the trade union.

2001 to 2006

Following the global stock market based dot com crash, the group undertook a major board restructuring and asset sale to address its debt pile. In May 2001, BT announced a three for ten rights issue to raise £5.9 billion – still the UK's largest ever rights issue – and the sale of Yell Group, the international directories and associated e-commerce business for £2.14 billion. Both activities were completed in June 2001. The group also sold its property portfolio to JV property company Telereal.

In April 2003, BT unveiled its current corporate identity and brand values. Reflecting the aspirations of a technologically innovative future, the connected world is designed to embody BT's five corporate values: trustworthy, helpful, inspiring, straightforward, heart.

The Communications Act, 2003 which came into force on 25 July 2003 introduced a new industry regulator, the Office of Communications (Ofcom), to replace the Office of Telecommunications (Oftel). It also introduced a new regulatory framework. The licensing regime was replaced by a general authorisation for companies to provide telecommunications services subject to general conditions of entitlement and, in some instances, specific conditions. Under a specific condition BT retained its universal service obligation (USO) for the UK, excluding the Hull area. The USO included connecting consumers to the fixed telephone network, schemes for consumers with special social needs, and the provision of call box services.

In the summer of 2004, BT launched Consult 21, an industry consultation for BT's 21st century network (21CN) programme. 21CN is a next generation network transformation, that, at one time, was due for completion by the end of 2010. Using internet protocol technology, 21CN will replace the existing networks and communications from any device such as mobile phone, PC, PDA or home phone, to any other device.

Openreach

Following the Telecommunications Strategic Review (TSR), in September 2005 BT signed legally binding Undertakings with Ofcom to help create a new regulatory framework for BT and the UK telecoms industry generally.

Openreach provides provision and repair in the "last mile" of copper wire. Formed from 25,000 engineers previously employed by BT's Retail and Wholesale divisions. It is designed to ensure that other communications providers (CPs) have exactly the same operational conditions as parts of the BT group. Opened for business on 11 January 2006, it reports directly into the BT chief executive.[15]

An Openreach van.

Acquisitions

In 2005, BT made a number of important acquisitions.

In February 2005, BT acquired El Segundo, California-based telecoms giant Infonet (now re-branded BT Infonet), giving BT access to new geographies.

It also acquired the second largest telecoms operator in the Italian business market, Albacom.

In April 2005, it bought Radianz from Reuters (now rebranded as BT Radianz), which expanded BT's coverage, provided BT with more buying power in certain countries.[16]

2006 to present

In August 2006 BT acquired online electrical retailer Dabs.com for £30.6 million.[17] The BT Home Hub manufactured by Inventel was also launched in June 2006.[18] In October 2006 BT confirmed that it would be investing 75% of its total capital spending, put at £10 billion over five years, in its new Internet Protocol (IP) based 21st century network (21CN). Annual savings of £1 billion per annum were expected when the transition to the new network was to have been completed in 2010, with over 50% of its customers to have been transferred by 2008. (For actual progress see BT 21CN). That month the first customers on to 21CN was successfully tested at Adastral Park in Suffolk.[19]

In January 2007, BT acquired Sheffield based ISP, PlusNet plc, adding an additional 200,000 customers. BT stated that PlusNet will continue to operate separately out of its Sheffield head-office.[20] On 1 February 2007 BT announced agreed terms to acquire International Network Services Inc. ("INS"), an international provider of IT consultancy and software. This increases BT presence in North America enhancing BT's consulting capabilities.[21] On 20 February 2007 Sir Michael Rake, then chairman of accountancy firm KPMG International, would succeed Sir Christopher Bland, who stepped down in September of that year.[22] On 20 April 2007 BT acquired Comsat International which provides network services to the South American corporate market.[23] On 1 October 2007 BT purchased Chesterfield based Lynx Technology which has been around since 1973. [24]

BT acquired Wire One Communications in June 2008 and folded the company into BT Conferencing, its existing conferencing unit, as a new video business unit[25] In July 2008, BT acquired the online business directory firm Ufindus for £20 million in order to expand its position in the local information market in the UK.[26] On 28 July 2008, BT acquired Ribbit, of Mountain View, California, "Silicon Valley's First Phone Company." Ribbit provides Adobe Flash/Flex APIs, allowing web developers to incorporate telephony features into their software as a service (SaaS) applications.[27]

On 1 April 2009 BT Engage IT is created from the unclean merger of two previous BT acquisitions Lynx Technology and Basilica. Apart from the name change not much else changes in operations for another 12 months. [28] On 14 May 2009 BT said it was cutting up to 15,000 jobs in the coming year after it announced its results for the year to 31 March 2009.[29] In July 2009 BT offered workers a long holiday for an up front sum of 25% of their annual wage or a one-off payment of £1000 if they agree to go part-time.[30]

Operations

British Telecommunications plc (BT) is a wholly owned subsidiary of BT Group plc and encompasses virtually all businesses and assets of the BT Group.[31] BT Group plc is listed on stock exchanges in London and New York.

The Adastral Park campus at Martlesham Heath in Suffolk, the principal site of BT Research.

BT runs the telephone exchanges, trunk network and local loop connections for the vast majority of British fixed-line telephones. Currently BT is responsible for approximately 28 million telephone lines in the UK. Apart from Kingston Communications, which serves Kingston upon Hull, BT is the only UK telecoms operator to have a *Universal Service Obligation* (USO) which means it must provide a fixed telephone line to any address in the UK. It is also obliged to provide public call boxes.

BT's businesses are operated under special government regulation by the British telecoms regulator Ofcom (formerly Oftel). BT has been found to have Significant Market Power in some markets following Market Reviews by Ofcom. In these markets, BT is required to comply with additional obligations such as meeting reasonable requests to supply services and not to discriminate.

As well as continuing to provide service in those traditional areas in which BT has an obligation to provide services or is closely regulated, BT has expanded into more profitable products and services where there is less regulation. These are principally, broadband internet service and bespoke solutions in telecommunications and information technology.

Divisions

BT Group is organised into the following divisions:[32]

- Customer facing:
  - BT Global Services - provides IT and telecoms services to multinationals (formerly **BT Ignite** and **BT Syntegra**)
  - BT Retail - provides retail telecoms services to businesses and consumers
  - BT Wholesale - operates BT's networks
  - Openreach - fenced-off wholesale division, responsible for the 'last mile' of BT's access network in the UK and tasked with ensuring that rival operators have equality of access to BT's local network
- Internal service units:
  - **BT Innovate and Design** - pulled together IT designers from BT Retail, BT Wholesale, BT Global Services and OneIT to design services on the 21C network.[33]
  - **BT Operate** - responsible for the roll-out and maintenance of BT's new IP based fixed-line network, known as 21st Century Network (21C)

Headquarters

BT Group's world headquarters and registered office is the BT Centre, a 10-storey office building at 81 Newgate Street in the City of London, opposite St. Paul's tube station.

Corporate affairs

Financial performance

Year ending	Turnover (£m)	Profit/(loss) before tax (£m)	Net profit/(loss) (£m)	Basic eps (p)
31 March 2011	20,076	1,717	1,504	19.4
31 March 2010	20,911	1,007	1,029	13.3
31 March 2009	21,390	(134)	(81)	3.2
31 March 2008	20,704	1,976	1,738	21.5
31 March 2007	20,223	2,484	2,852	34.4
31 March 2006	19,514	2,633	1,644	19.5
31 March 2005	18,429	2,693	1,539	18.1
31 March 2004	18,519	1,945	1,414	16.4
31 March 2003	18,727	3,157	2,702	31.4
31 March 2002	18,447	1,461	1,008	12.1
31 March 2001	17,141	(1,031)	(1,875)	(25.8)
31 March 2000	18,715	2,942	2,055	31.7
31 March 1999	16,953	4,295	2,983	46.3
31 March 1998	15,640	3,214	1,702	26.6
31 March 1997	14,935	3,203	2,077	32.8
31 March 1996	14,446	3,019	1,986	31.6
31 March 1995	13,893	2,662	1,731	27.8
31 March 1994	13,675	2,756	1,767	28.5
31 March 1993	13,242	1,972	1,220	19.8
31 March 1992	13,337	3,073	2,044	33.2

Pension fund

BT has the largest defined benefit pension plan of any UK public company. The trustees valued the scheme at £36.7 billion at the end of 2010;[34] An actuarial valuation valued the deficit of the scheme at £9.043 billion as of 31 December 2008.[35] However following a change in the regulations governing inflation index linking, the deficit was estimated at £5.2 billion in November 2010.[36] BT and the Trustee have agreed a 17 year recovery plan with the first three years' payments amounting to £525mm. As of 2013 average annual payments have been estimated at £533M. The next triennial valuation is scheduled for December 2011.[36]

BT's pension obligation is derived from two pension plans: BTPS, the company's defined-benefit pension scheme which was closed in 2001, and the BT Retirement Saving Scheme (BTRSS), which was set up to replace the BTPS and is a defined-contribution retirement plan.

Logo

The current BT corporate logo, "Connected World", was adopted in 2003.[37] The globe device part of the logo was originally designed by the Wolff Olins brand consultancy for BT's Concert joint venture with AT&T, and was subsequently used by BT's internet division, Openworld, prior to being adopted by the company as a whole.[37]

1980 – 1991
British Telecom "T" symbol logo.

1991 – 2003
BT "piper" logo.

2003 – present
BT "Connected World" logo.

Environmental record

In 2004, the BT Group signed the world's largest renewable energy deal with npower and British Gas, and now all of their exchanges, satellite networks and offices are powered by renewable energy. BT is a member of the Corporate Leaders Group on Climate Change. They signed a letter urging the government to do more to tackle this problem. Janet Blake, head of global corporate social responsibility (CSR) at BT, says that she would like to see incentives that find ways of rewarding those companies that focus on climate change by making investments in green business models.[38]

BT has made it clear that it has an ambitious plan to reduce carbon dioxide emissions.[39] Its strategy includes steps to reduce the company's carbon footprint as well as those of customers, suppliers and employees. BT has actually pledged to achieve an 80% reduction by the year 2016, which will require further efficiency improvements.[40]

Since June 2009 BT has been one of the founding members of the Ellen MacArthur Foundation.

Controversies

World Wide Web hyperlink patent

In 2001, BT discovered it owned a patent (U.S. Patent 4873662 [41]) which it believed gave it patent rights on the use of hyperlink technology on the World Wide Web. The corresponding UK patent had already expired, but the US patent was valid until 2006. Opponents of BT's claim held that the patent had never been valid, due to prior art by both Douglas Engelbart and Ted Nelson's Project Xanadu. Nevertheless, on 11 February 2002, BT began a court case relating to its claims in a U.S. federal court against the Internet service provider Prodigy Communications Corporation. In the case *British Telecommunications Plc. v. Prodigy*, the United States District Court for the Southern District of New York ruled on 22 August 2002 that the BT patent was not applicable to web technology and granted Prodigy's request for summary judgement of non-infringement.[42] The issue of prior art was thus not addressed.

Behavioural targeting

In early 2008 it was announced that BT had entered into a contract (along with Virgin Media and TalkTalk) with the spyware company Phorm (responsible under their 121Media guise for the Apropos rootkit)[43] [44] to intercept and analyse their users' click-stream data and sell the anonymised aggregate information as part of Phorm's OIX advertising service.[45] [46] The practice, known as "behavioural targeting" and condemned by critics as "data pimping", came under intense fire from various internet communities and other interested-parties who believe that the interception of data without the consent of users and web site owners is illegal under UK law (RIPA).[47] [48] [49] [50] At a more fundamental level, many have argued that the ISPs and Phorm have no right to sell a commodity (a user's data, and the copyright content of web sites) to which they have no claim of ownership. In response to questions about Phorm and the interception of data by the Webwise system Sir Tim Berners-Lee, credited as the creator of the World Wide Web, indicated his disapproval of the concept and is quoted as saying of his data and web history:

> It's mine – you can't have it. If you want to use it for something, then you have to negotiate with me. I have to agree, I have to understand what I'm getting in return. I myself feel that it is very important that my ISP supplies internet to my house like the water company supplies water to my house. It supplies connectivity with no strings attached. My ISP doesn't control which websites I go to, it doesn't monitor which websites I go to.
>
> — Sir Tim Berners-Lee, 2008[51]

References

[1] http://www.londonstockexchange.com/exchange/prices-and-news/stocks/prices-search/stock-prices-search.html?nameCode=BT.A
[2] http://www.nyse.com/about/listed/quickquote.html?ticker=bt
[3] "BT Annual Report and Form 20– F 2012" (http://www.btplc.com/Sharesandperformance/Annualreportandreview/pdf/BTAnnualReport2012_smart.pdf). BT Group plc. . Retrieved 27 May 2012.
[4] http://www.btplc.com/
[5] "Our company" (http://www.btplc.com/Thegroup/Ourcompany/index.htm). BT Group plc. Archived (http://web.archive.org/web/20100829014043/http://www.btplc.com/thegroup/ourcompany/index.htm) from the original on 29 August 2010. . Retrieved 24 August 2010.
[6] "BT Global Services" (http://www.btplc.com/Thegroup/Ourcompany/Companyprofile/Groupbusinesses/BTGlobalServices/index.htm). BT Group plc. Archived (http://web.archive.org/web/20100810172518/http://www.btplc.com/Thegroup/Ourcompany/Companyprofile/Groupbusinesses/BTGlobalServices/index.htm) from the original on 10 August 2010. . Retrieved 24 August 2010.
[7] "BT Retail" (http://www.btplc.com/Thegroup/Ourcompany/Companyprofile/Groupbusinesses/BTRetail/index.htm). BT Group plc. . Retrieved 24 August 2010.
[8] "Lifetime Legacy" (http://www.globalservices.bt.com/uk/en/insights/lifetime_legacy). . Retrieved 18 December 2012.
[9] "BT takes on challenge of London's hyperconnected Olympic Games" (http://www.totaltele.com/view.aspx?ID=474863). . Retrieved 18 December 2012.
[10] "FTSE All-Share Index Ranking" (http://www.stockchallenge.co.uk/ftse.php). stockchallenge.co.uk. . Retrieved 26 December 2011.
[11] "The historical development of BT" (http://www.btplc.com/Thegroup/BTsHistory/History.htm). BT Group. 2010. Archived (http://web.archive.org/web/20100527143141/http://www.btplc.com/Thegroup/BTsHistory/History.htm) from the original on 27 May 2010. . Retrieved 12 June 2010.
[12] Tory conference bombed (http://www.expressandstar.com/days/1976-2000/1984.html) Express and Star. Retrieved 24 March 2012
[13] "BT gobbles up Cellnet" (http://news.bbc.co.uk/1/hi/business/404792.stm). BBC News. 27 July 1999. . Retrieved 1 May 2010.
[14] "BT approves mobiles spin-off" (http://news.bbc.co.uk/1/hi/business/1615100.stm). BBC News. 23 October 2001. . Retrieved 1 April 2010.
[15] BT's Openreach could have to list separately (http://www.allbusiness.com/media-telecommunications/8208794-1.html)
[16] "BT goes home" (http://findarticles.com/p/articles/mi_m0DUJ/is_/ai_n9772149). Findarticles.com. 1 December 2004. . Retrieved 18 April 2011.
[17] "BT buys on-line retailer Dabs.com" (http://news.bbc.co.uk/1/hi/business/4956100.stm). BBC News. 28 April 2006. . Retrieved 18 April 2011.
[18] "Thomson ships BT home hub" (http://www.lightreading.com/document.asp?doc_id=105546). Lightreading.com. . Retrieved 18 April 2011.
[19] "BT transfers first customer lines to 21CN network" (http://www.highbeam.com/doc/1G1-155608023.html). Highbeam.com. 1 December 2006. . Retrieved 18 April 2011.
[20] Thurston, Richard (13 November 2008). "Bt buys PlusNet for CRM system" (http://news.zdnet.co.uk/communications/0,1000000085,39284743,00.htm). News.zdnet.co.uk. . Retrieved 18 April 2011.
[21] "BT buys former Lucent division INS" (http://www.computerwire.com/industries/research/?pid=6E892FE0-2D6A-4870-82DA-2EFD22D6D788). Computerwire.com. . Retrieved 18 April 2011.
[22] "Datamonitor ComputerWire – BT Appoints New Chairman to Replace Bland" (http://www.computerwire.com/industries/research/?pid=FFF73A29-8EAC-4FA6-BD80-883AD44E3CCB). Computerwire.com. . Retrieved 18 April 2011.
[23] "BT buys Comsat" (http://telephonyonline.com/finance/news/bt_buys_comcast_042007/). Telephonyonline.com. . Retrieved 18 April 2011.

[24] Meyer, David (3 October 2007). "BT buys Lynx Technology" (http://www.zdnet.co.uk/news/systems-management/2007/10/01/bt-buys-lynx-technology-39289743/). Zdnet.co.uk. . Retrieved 18 April 2011.

[25] BT buys Wire One (http://www.vnunet.com/vnunet/news/2214152/bt-acquires-wire-one)

[26] Wray, Richard (10 July 2008). "BT buys into social networking directories with £20m for Ufindus" (http://www.guardian.co.uk/business/2008/jul/10/btgroupbusiness.socialnetworking). *The Guardian* (London). .

[27] "BT buys Ribbit for $105m" (http://gigaom.com/2008/07/29/bt-buys-ribbit/). Gigaom.com. 29 July 2008. Archived (http://web.archive.org/web/20110515065153/http://gigaom.com/2008/07/29/bt-buys-ribbit/) from the original on 15 May 2011. . Retrieved 18 April 2011.

[28] Exclusive: Lynx and Basilica in BT rebrand (http://www.channelweb.co.uk/crn-uk/news/1892864/exclusive-lynx-basilica-bt-rebrand)

[29] BT to shed a further 15,000 job losses (http://news.bbc.co.uk/1/hi/business/8049276.stm) BBC News, 14 May 2009

[30] "Bbc News" (http://news.bbc.co.uk/1/hi/business/8134123.stm). BBC News. 4 July 2009. . Retrieved 18 April 2011.

[31] "2008 Annual Report Overview" (http://www.btplc.com/report/report08/overview/index.htm). .

[32] "Group businesses" (http://www.btplc.com/Thegroup/Ourcompany/Groupbusinesses/index.htm). BT Group plc. . Retrieved 22 February 2012.

[33] "Two new BT businesses after a radical shake-up". *Financial Times*. 25 April 2007.

[34] BT Pension scheme Report and Accounts (http://www.btpensions.net/41/scheme-report-and-accounts)

[35] "BT Pension Scheme Valuation 2008" (http://www.cwu-cctv.org/article.php?articleid=259). . Retrieved 4 August 2011.

[36] "BT pension deficit falls £2.9bn on CPI inflation link change" (http://uk.finance.yahoo.com/news/bt-pension-deficit-falls-2-9bn-on-cpi-inflation-link-change-tele-4c5fdd6acc90.html). *The Daily Telegraph*. UK. . Retrieved 14 March 2011.

[37] "BT calls up a new logo in attempt to escape from 'arrogant' image" (http://www.independent.co.uk/news/business/news/bt-calls-up-a-new-logo-in-attempt-to-escape-from-arrogant-image-593715.html). *The Independent* (UK). 8 April 2003. . Retrieved 25 January 2012.

[38] Interview: CSR delivers £2.2bn for BT – 30 October 2006 – IT Week (http://www.itweek.co.uk/business-green/analysis/2199869/interview-csr-delivers-2bn-bt)

[39] "Climate change" (http://www.btplc.com/Societyandenvironment/Climatechange/Climatechange.htm). BT Group. . Retrieved 18 April 2011.

[40] "Climate change – BT's Vision and Strategy" (http://www.btplc.com/ClimateChange/WhatsBTdoing/Visionandstrategy/index.htm). BT Group. . Retrieved 18 April 2011.

[41] http://www.google.com/patents?vid=4873662

[42] "BT's "Hyperlinking" Patent Litigation Fails" (http://www.nswscl.org.au/journal/51/Glen_Sauer.html). Nswscl.org.au. . Retrieved 18 April 2011.

[43] F-Secure Spyware Information Pages: Apropos (http://www.f-secure.com/sw-desc/apropos.shtml)

[44] "F-Secure Spyware Information Pages: PeopleOnPage" (http://www.f-secure.com/sw-desc/peopleonpage.shtml). F-secure.com. Archived (http://web.archive.org/web/20110524153535/http://www.f-secure.com/sw-desc/peopleonpage.shtml) from the original on 24 May 2011. . Retrieved 18 April 2011.

[45] Williams, Chris (25 February 2008). "ISP data deal with former 'spyware' boss triggers privacy fears" (http://www.theregister.co.uk/2008/02/25/phorm_isp_advertising/page2.html). *The Register*. . Retrieved 18 April 2011.

[46] Williams, Chris (29 February 2008). "How Phorm plans to tap your internet connection" (http://www.theregister.co.uk/2008/02/29/phorm_documents/). *The Register*. . Retrieved 18 April 2011.

[47] "Web users angry at ISPs' spyware tie-up" (http://www.thisislondon.co.uk/standard-home/article-23449601-details/Web+users+angry+at+ISPs'+spyware+tie-up/article.do). *Evening Standard*. London. . Retrieved 18 April 2011.

[48] Williams, Chris (4 March 2008). "Data pimping: surveillance expert raises illegal wiretap worries" (http://www.theregister.co.uk/2008/03/04/phorm_ripa/). *The Register*. . Retrieved 18 April 2011.

[49] Williams, Chris (17 March 2008). "Net think thank: Phorm is illegal" (http://www.theregister.co.uk/2008/03/17/phorm_fipr_illegal/). *The Register*. . Retrieved 18 April 2011.

[50] "The Phorm "Webwise" System – a Legal Analysis" (http://www.fipr.org/080423phormlegal.pdf) (PDF). Archived (http://web.archive.org/web/20110522062819/http://www.fipr.org/080423phormlegal.pdf) from the original on 22 May 2011. . Retrieved 18 April 2011.

[51] Cellan-Jones, Rory (17 March 2008). "Web creator rejects net tracking" (http://news.bbc.co.uk/1/hi/technology/7299875.stm). BBC News. . Retrieved 18 April 2011.

External links

- BT Group (http://www.btplc.com/)
- BBC news story on BT's claimed 'web patent' (http://news.bbc.co.uk/hi/english/sci/tech/newsid_798000/798475.stm)
- BT's patent text at USPTO (http://www.google.com/patents?vid=4873662)
- Patent case court judgement against BT (http://www.nysd.uscourts.gov/courtweb/pdf/D02NYSC/02-07733.PDF)

Article Sources and Contributors

Viscount_Hanworth *Source*: http://en.wikipedia.org/w/index.php?title=Viscount_Hanworth *Contributors*: Adam37, Alci12, Berks105, Ipankonin, Jordi22, Lord Emsworth, Rrius, Scwlong, Tryde, 3 anonymous edits

Tony_Benn *Source*: http://en.wikipedia.org/w/index.php?title=Tony_Benn *Contributors*: !!, 1exec1, A bit iffy, AJMW, Abdelhk, Academic Challenger, Adam Carr, Adam37, Afterwriting, Alan Peakall, Alci12, All Hallow's Wraith, AllanHainey, Alligators1974, AndySimpson, Anglicanus, Anna Frodesiak, Apeloverage, Ashley Pomeroy, Atchom, Athenianvoice, Avaya1, Avianflu, Avicennasis, AxelBoldt, Azerbaijanman, BMF81, Badgerpatrol, Bagworm, Bahauddeen, Bazj, Bean is a carrot, Bearcat, Bedivere, Beetstra, Ben davison, Bender235, Berks105, Bgh251f2, Bill Tegner, Bilsonius, Biruitorul, Bletchley, Blorg, Bobo12345, Bodnotbod, Booklung, Bradley0110, Breadandcheese, Brian McNeil, Britannicus, BritishWatcher, BrownHairedGirl, Burzmali, C.bennetts, CCorward, CPMcE, Canistabbats, Catchpole, Causa sui, Cavie78, Chicheley, Child of Albion, Chomsky123, Chris j wood, Chris the speller, Chrisjsaunders, Chuq, Clivemacd, Clutchy, Colonies Chris, Combating Ignorance, Craverguy, Croctotheface, D6, Dahn, Danbarnesdavies, Daniel.finnan, DanielRigal, Davewild, Deb, Deipnosophista, Delirium, Deltabeignet, Delysid, Demiurge, Deor, Destynova, Deville, Diodecimus, Djnjwd, Dmn, Docu, Domminico, Donation, DrFrench, Duffy2032, DuncanHill, Ed g2s, Edgar181, Edward, Eloquence, Erp, Eugenespeed, Fenster, Finlay McWalter, Fragglet, Franklinville, Fys, GeeJo, Georgereev118118, Gingerblokey, GoldenTorc, Good Olfactory, GreatToBe18, Greenshed, Ground Zero, Grutness, Guyalsfere, Hairy Dude, Haldraper, Hammersfan, HampshireCricketFan, Happywith006, Hauskalainen, Heartylunch, Henry Flower, Histmag, Hoary, Hunkeemunkee, Hux, Ipankonin, Irishpunktom, Iron Ghost, J.delanoy, JJARichardson, JJE, JK the unwise, JRawle, JackofOz, Jamesgibbon, Jamesmorrison, Jatkins, Jb849, Jdforrester, Jellyman, Jerlam2, Jetman, Jezhotwells, Jienum, Jim Michael, Joffan, John, John K, John21Allen, Johncmullen1960, Jongleur100, Jpbowen, Jxm, Kabads, Kaihsu, KarlJorgensen, Katherine Shaw, Kbdank71, Keith D, Keith Edkins, Koavf, Kolsteinz, Kuralyov, LAWinans, LJ Holden, Lapsed Pacifist, Larklight, Lawrence Waterhouse, Lawsonrob, Lee M, Lifthrasir1, Lomcevak, Longlivefolkmusic, Lord Emsworth, Lord.lucan, Lr676r, Lrty67, Lucian Sunday, MacGonagall, MacRusgail, Maelnuneb, Magic Lemur, Malarious, Malik Shabazz, Marian Silvester, MarkThomas, Marktreut, MaybeMaybeMaybe, Mendors, Mervyn, Midnightblueowl, MikeMCSG, Mikedelsol, Mintguy, Mobilecinema, Mobit, Montrealais, Monyvibescu, Moonriddengirl, Morwen, Mountdrayton, Mpntod, Mrrhum, Mrsteed, Ms454, Mtaylor848, Mtiedemann, Muckna, Mutt, Nandt1, NathanPaulBridle, Necrothesp, Neilbeach, Neutrality, Nick Smale, Nommonomanac, Not050, Obstantia, Ohconfucius, OldakQuill, Olebin, One Night In Hackney, Organic Cabbage, Ou tis, PBS, Paj.meister, PatGallacher, Paul Barlow, Paul MacDermott, Paul W, PaulS89, Pcpcpc, Pgan002, Phenry99, Philip Cross, Philip Stevens, Pigsonthewing, Pkchan, Pointer1, Profoss, Proteus, Ps75, Pterre, Qwertyus, RFBailey, RJTinsley, Rachel Pearce, RafikiSykes, Rbrwr, Rd232, Redthoreau, Rekab Mit, Riadismet, Rich Farmbrough, Rjwilmsi, Rms125a@hotmail.com, Road Wizard, RobStreatham, Robert Saunders, RodCrosby, RolandR, Rrius, Russpear, Rwendland, SDC, SE7, ST47, Saluton, Sam Blacketer, Saulisagenius, SchuminWeb, Scott MacDonald, Sealman, Secretlondon, Seth Whales, ShelfSkewed, Sherefong, Shfletcher, Shirelord, Shortenfs, SimonD, Sir Stanley, Sjakkalle, Sjorford, Slash, Socialdemocrats, Spicemix, SqueakBox, Squiquifox, Srheller, Stebbins, Steinsky, Stephenb, Stephennarmstrong, Superm401, Swangyy, Tagishsimon, Taigkiller@earthnet.link, Tamas, Tamfang, Terryitaloan, The Missing Hour, Therequiembellishere, Tiles, Tim Ivorson, Tim!, Timrollpickering, TomyDuby, Tony Corsini, Tony1, Tonywalton, Torqueing, Tpbradbury, Tubalib, UESPArules, UKWiki, UpDown, Vaganyik, Valentinejoesmith, Victuallers, Vis-a-visconti, WTD, Walker44, Wallie, Warofdreams, Wedgwoodpound, WereSpielChequers, West Brom 4ever, WikianJim, Woohookitty, Yamamoto Ichiro, Zadcat, 516 anonymous edits

Charles_Longley *Source*: http://en.wikipedia.org/w/index.php?title=Charles_Longley *Contributors*: 1549bcp, Access Denied, Adam sk, Amalas, Anglicanus, Arla, D6, Danbarnesdavies, Dcoetzee, Deb, DragonflySixtyseven, Dsp13, Egham1, Gadfium, GregorB, Grenavitar, Jaraalbe, Jdforrester, Jpbowen, Keith D, Keith1234, Malleus Fatuorum, MarmadukePercy, Mcc2inf, Mcevjp, Mikeindex, Moonraker, Neddyseagoon, Notjohn, Ohconfucius, PMJ, Plastikspork, Plucas58, Ptolemy Caesarion, Regan123, Rich Farmbrough, SilkTork, The Duke of Waltham, Tim!, Timrollpickering, Utcursch, Waacstats, Wikibiohistory, Yworo, 13 anonymous edits

John_Thurso *Source*: http://en.wikipedia.org/w/index.php?title=John_Thurso *Contributors*: Ablevive, Alai, Anbu121, BasilSorbie, Billydeeuk, Biscit, BlueTemplar12, BrownHairedGirl, ChrisGualtieri, Conti, Craigy144, D6, DWaterson, David.Mestel, Delirium, Flaming Ferrari, Flatterworld, Gary J, Ground Zero, H2g2bob, Henrygb, Ipankonin, JRawle, Jdforrester, Kbsmarion, Keith Edkins, Kernel Saunters, Laurel Bush, Law Lord, Lord Emsworth, Mabelina, Mais oui!, Malcolmxl5, Man vyi, Matthew Platts, Morwen, Ninetyone, Nogwa, OutragedOfOake, Peetoor, Philip Stevens, Rbrwr, Re10cc66, Red stucco, Rje, Rrius, Russhayley, SE7, Sam Blacketer, Sebmelmoth, Shortride, SimonArlott, Sky4t0k, SoM, Spellmaster, Stephennarmstrong, TEB728, TheParanoidOne, Timrollpickering, Tryde, Ugen64, Uncantabrigian, WOSlinker, Wereon, Widefox, Wik, Wizardman, Woohookitty, X76trombones, Zeth, 45 anonymous edits

Douglas_Hogg *Source*: http://en.wikipedia.org/w/index.php?title=Douglas_Hogg *Contributors*: Alci12, Ausseagull, Average Earthman, Badgerpatrol, Barticus88, Berks105, Bilsonius, Brandon97, BritishWatcher, BrownHairedGirl, Cannonmc, Charles01, Charlotte Leslie 1978, Christopherlin, ContiAWB, Craigy144, D6, David Gerard, David the Aspie, Dcrossle, DinosaursLoveExistence, Dondilly, Dottydotdot, Duffy2032, Ericoides, FedLawyer, Flaming Ferrari, Frailea, Fys, G Colyer, Genisock2, George Burgess, GrahamSmith, Hackloon, Hammersfan, Histmag, Ipankonin, Ivor Stoughton, JRRobinson, JRawle, JackHearne, Jdforrester, John, Jsek, Keith D, Keith-264, Kittybrewster, Lincolnite, Linksjinx, Lord Emsworth, Mackensen, Messenger88, Michael Drew, Mintysmum, Mootros, Morwen, Mtiedemann, Neelix, Newprogressive, Obeyyourmind, Opera hat, Pauly04, Philip Cross, Philip Stevens, Phoe, Proteus, Qp10qp, Reggie999, Rjwilmsi, Rlarrington, Rms125a@hotmail.com, Road Wizard, RobStreatham, Ronald Collinson, Rossybeary, Rrius, Rwendland, SE7, SMcCandlish, Saga City, Salvio giuliano, Samuel Salzman, SimonP, Smerus, SqueakBox, The Duke of Waltham, TheParanoidOne, Thegn, Therequiembellishere, Timrollpickering, Timtrent, Trident13, Ugen64, Uncle Dick, ViennaUK, Wally, Welsh, Wereon, William Avery, Yorkshirian, Zaphod Beeblebrox, Zviki1, 82 anonymous edits

Viscount_Cobham *Source*: http://en.wikipedia.org/w/index.php?title=Viscount_Cobham *Contributors*: Aimsworthy, Alci12, Berks105, Colourman, CommonsDelinker, Craigy144, DanMS, Fayenatic london, Heavens To Betsy, Henning M, Histmag, Ipankonin, Jdforrester, Laggan Boy, Lord Emsworth, M2o6n0k2e7y3, MER-C, Mean as custard, Merovingian, Motmit, Opera hat, PBS, Phoe, Pilch62, Proteus, Rjm at sleepers, Rrius, SDC, Scwlong, Tamfang, Tim riley, Timrollpickering, Tryde, Wjhonson, 7 anonymous edits

Viscount_Chandos *Source*: http://en.wikipedia.org/w/index.php?title=Viscount_Chandos *Contributors*: Before My Ken, Berks105, El estremeñu, FeralHamster, Henning M, Histmag, Ipankonin, Jdforrester, Jeni, Lord Emsworth, Merovingian, Miami33139, Old Moonraker, Phoe, Rrius, Scwlong, Tryde, Winhunter, YUL89YYZ, 8 anonymous edits

Liberal_Party_(UK) *Source*: http://en.wikipedia.org/w/index.php?title=Liberal_Party_%28UK%29 *Contributors*: 130.94.122.xxx, Adam Carr, Adashiel, Al B. Free, Alexanderps, Andrew Dalby, Andrewblagger, Angela91210, Aristotle, AsaRoast, Autospark, Ayrshire95, Balcer, Bastin, Batmanand, Bkwillwm, Blorg, Bob bobato, Bobblewik, Boris Kaiser, Britannicus, BrownHairedGirl, Byelf2007, Calabe1992, Cartwright27, ChancellorAlbatross, Checco, Chekaz, Chrisieboy, Colonies Chris, Conscious, Conversion script, Coyets, Craigy144, Cremepuff222, DaQuirin, Darolew, DerHexer, Derek Ross, Dermo69, Dn9ahx, Elmondo21st, Epbr123, Erath, Flopster2, Formeruser-83, Freedomlinux, Fys, G-Man, Gabbe, Gaius Cornelius, Gary J, Gerard Mulholland, GoodDay, GorillaWarfare, Gr8opinionater, Graham Lippiatt, Ground Zero, Haldraper, Handy249, Henrygb, Hephaestos, Ikanreed, Indrian, Isaacwatts, J.delanoy, JJ Williams, JMonroe, JW1805, JaT, Jdforrester, JeremyA, Jnestorius, Joelwhit, John K, JohnI, Joseph Solis in Australia, Jttw, JustAGal, Kanguole, Kendrick7, Kglavin, Kjetil82, Kross, Laurinavicius, Levg, Lfuyt, Lockesdonkey, Lofty, Longlivefolkmusic, Mackensen, Magioladitis, Marcus22, Mark Boyle, MartinezMD, Matthew Kilburn, MattyMei78, Mav, Maximus Rex, Meco, Mfcayley, Michael Hardy, Mintguy, Moe Epsilon, Money bagess, Morwen, Mtaylor848, Mátyás, Nicke L, Nicke Lilltroll, Nicklse, Niteowlneils, Nk, Osioni, PBS, Paul-L, Philip Stevens, Plastikspork, Proteus, Psychonaut, Purple1342, Pyrostalk, RFBailey, Rbrwr, Rcbutcher, Rdeacon, Reconsider the static, RekishiEJ, Richard David Ramsey, Rjensen, Road Wizard, Rossenglish, Rothorpe, Rrius, SE7, Saga City, SilkTork, Sir Stanley, Sjorford, Sky4t0k, Smjg, Sophie means wisdom, Spartiate, Squiddy, Stephen C. Carlson, Stevepday, Str1977, Stratercastguy, Sun King, Tartan, Teneriff, Texas.veggie, The Arbiter, The Deviant, The Proffesor, The Thing That Should Not Be, Tim!, Timrollpickering, Tony1, Toussaint, Twilde, Tyronen, UberCryxic, Useight, VampWillow, Viking880, VodkaJazz, Vrenator, Wapcaplet, Warofdreams, Wereon, WikHead, Wilfried Derksen, Woodgreener, Yohan euan o4, Zangar, 180 anonymous edits

Social_Democratic_Party_(UK) *Source*: http://en.wikipedia.org/w/index.php?title=Social_Democratic_Party_%28UK%29 *Contributors*: 130.94.122.xxx, 1exec1, 87h23BS87g, Andre Engels, Antandrus, Anti-socialism, Autospark, BScar23625, BartBassist, Bcj11, Beao, Bevo74, Biruitorul, Blovebloveb, Bollar, BrownHairedGirl, Buttyboy, CALR, CJ Aikman, Camerong, CarolGray, CiteSources, ColinHogben, CommonsDelinker, Conversion script, Danio, Bibliophylax, DaveK@BTC, Davewild, David Kernow, Davidcannon, Dirgela, Dl2000, Dogaroon, Duffy2032, Dupont Circle, Duuude007, Edward, EdwinHJ, Electrobe, Epbr123, Erath, Esperant, Estlandia, Everyking, FF2010, Flagboy, Fys, G-Man, Gaius Cornelius, Galloglass, Geni, Good Olfactory, Gr8opinionater, Grackelstein, Graham Lippiatt, Grarg, Greyisthenewblue, Ground Zero, GullibleKit, H2g2bob, Haldraper, Hazhk, Histmag, JackHearne, Jamesparkin, Jiang, John K, JohnI, Kalimac, Khazar, Kristallstadt, Kuralyov, Lawrence142002, Lawsonrob, Lightmouse, Lockesdonkey, Lord of the Isles, Lozleader, M-le-mot-dit, Mackensen, Marcus22, Mark Boyle, Meco, Medusa1991, Meiskam, Mfcayley, Mibblepedia, Midnightblueowl, Million Moments, Milo99, Mintguy, Mj wood, Morwen, Mpntod, MrWeeble, Multiculturalist, Myanw, Nach0king, Netsurfer123, Oliver Chettle, Orderinchaos, Ortcutt, PaddyBriggs, Palomar, PatGallacher, Philip Cross, Plastikspork, PleaseStand, Plinford, Psychonaut, Ptah, the El Daoud, Purple1342, Quiensabe, RJFJR, Rbrwr, Rcbutcher, Rich Farmbrough, Rjwilmsi, Roadrunner, RodCrosby, Rodhullandemu, Rossenglish, Rothorpe, Rrius, Ruy Lopez, SE7, Saltley, Sam Blacketer, Sgt Pinback, Shaunnol, Sheridan, Sir Stanley, Sjoh0050, Sky4t0k, SourcerRef, Srheller, Steinsky, Talrias, Tim!, Timrollpickering, Tuesdaily, Unschool, UtherSRG, Valenciano, VampWillow, Vclaw, Verrai, Viking880, Visitweak, Warofdreams, Widr, William Avery, Wizardman, Woodgreener, Yorkshirian, לבורמצא, 134 anonymous edits

Commonwealth *Source*: http://en.wikipedia.org/w/index.php?title=Commonwealth *Contributors*: $jonathannnnnnn, 16@r, 23prootie, Acegikmo1, Agemegos, Aitias, Alansohn, Alex.muller, AlisonW, Anaxial, Andrew Gray, Andrew Yong, Andy Marchbanks, AndySimpson, AnemoneProjectors, Anniika, Anomalocaris, Antandrus, Apmab1, Aprogressivist, ArCgon, ArnoLagrange, Arthena, Asdnhk, Ashleansmith, AxelHarvey, Aymatth2, Bcorr, Bemoeial, Bennó, Bentogoa, Big Adamsky, BilCat, Binabik80, Boomshadow, Brian the Editor, Buaidh, CJ, CL, Cabe6403, Calieber, Cameron, Carrionluggage, Cassowary, Chick Bowen, Chr1stian, Chrisvls, Chromenano, Claudine, Codingmasters, Collabi, ConCompS, Conversion script, Coolertyp, CopperSquare, Correogsk, Curb Chain, CyclePat, DVD R W, Daeron, Davewild, Davodude2, Decumanus, Deli nk, Diannaa, Diderot, Dietary Fiber, Discourseur, Docboat, Dolovis, Downtown dan seattle, Dreaded Walrus, Duffman, DuncanHill, Duoduoduo, Ed de Jonge, Egyptian lion, Emax, Eraserhead1, Eric-Wester, Etineskid, Fabartus, Faradayplank, Flowerpotman, Formeruser-81, Funnybunny, Gaius Cornelius, Gdr, Gentgeen, Gerald Farinas, Geraldo Perez, Gfoley4, Glacialfox, Glostos, GoldRingChip, Gritchka, Ground Zero, Grutness, Gswann06, Hadmack, Hairy Dude, Harjk, Hbis06r, Henrygb, Hetar, Huntington, IgnorantArmies, Iketsi, Improv, Isidore, Isis, J.delanoy, JAJ, JForget, JHMM13, JHunterJ, JNW, JPD, JW1805, Jack Upland, Jake Wartenberg,

Jarry1250, JavierMC, Jeff G., Jenmoa, Jeronimo, Jim1138, Jimp, Jleonau, Jncraton, Jomoal99, Jonearles, Jonemerson, Jooler, Jrockley, Jtdirl, Jusdafax, Kablammo, Kaihsu, Kaleal92, Kateshortforbob, Kesuari, Kevin Rector, KnowledgeOfSelf, Kpalion, Kurykh, Kyle sb, L1A1 FAL, Lacrimosus, LairepoNite, LanternLight, LedgendGamer, Lemonade100, LibreExObscur, Lnegro, Lorenzarius, Lupin, MER-C, MPD01605, MaartenVidal, Madhero88, Malcolm Farmer, Mallerd, Manapy, Mark83, Marketdiamond, Martinp23, Meiskam, Menchi, Merrymanhappy, Michael Hardy, Miesianiacal, Mikeh, Minesweeper, Miriam650611, Mogism, Molinari, Montell 74, Montrealais, Mr Gronk, Ndenison, Neomanxl, NewEnglandYankee, Nishkid64, Niteowlneils, Noremacmada, Nubiatech, Num1dgen, Numerousfalx, Ohnoitsjamie, Olivier, OpenInfoForAll, Optigan13, Oxymoron83, PalawanOz, Pasajero, PaulAllison, Pb30, Peterlewis, Pharos, PhilHibbs, Philip Trueman, Piano non troppo, Piotrus, Pne, Provocateur, Quiensabe, R'n'B, RJC, RackAttack, Radon210, RandomCritic, Recognizance, Reginald Perrin, Reginmund, Restraining, Roadrunner, Robin S, Robodoc.at, Rothorpe, Royptorico, SchfiftyThree, Semper discens, Sengkang, Shuggyg, SinghPz, Skellar, Smooth O, Solar765, Soulpatch, SpikeToronto, SpookyMulder, StAnselm, StaticGull, Stec, Stephen C. Carlson, Stuartjmoore, Sumirp, Sven Manguard, SyL64, TShilo12, Teles, The Man in Question, The Thing That Should Not Be, Thedreamdied, Thehelpfulone, Themoodyblue, Theone256, Thorwald, Thursiya, TigerShark, Tj57, Tombombadil, TreasuryTag, Uk34, UncleDouggie, Urhixidur, Vierutasd, Vipinhari, Vivio Testarossa, Vzbs34, WHEELER, WMC, Weedwhacker128, Werdna, WhisperToMe, Wik, Wikipelli, William M. Connolley, Woohookitty, Work permit, Zalgo, ZekMin, Zmaj, Zondor, 628 anonymous edits

Concorde *Source*: http://en.wikipedia.org/w/index.php?title=Concorde *Contributors*: -Majestic-, 1995hoo, 1exec1, 200.191.188.xxx, 334a, 50pence, 621PWC, 790, 84user, 9913, A strolling player, A. Parrot, A380 Fan, ACY.Man, AD66, AEMoreira042281, AHands, AMCKen, ASV47, Abut, Acb58, Adimage, Adry1991, AdultSwim, Aeroeditin, Afowler, Ahoerstemeier, Ahunt, Aircorn, Airodyssey, Airplaneman, Ajkgordon, Ajray, Akradecki, Alai, Alansohn, Alaz, Albany NY, Aleenf1, Alex756, Alfio, Ali@gwc.org.uk, Alias777, Alsandro, Alvaro, Alvis, Amitkinger, Amplitude101, Amthomas0412, An2597, Andrejj, Andresihera, Android Mouse, Andy Dingley, Andy G, AnthonyUK, Antonio Lopez, AntonioMartin, Archzog, Aremith, Armeria, Arpingstone, AscendedAnathema, Ask123, Askari Mark, Aspensti, Asybaris01, Atlant, Auseplot, Autair, Avaya1, AxG, Axver, Ayla, BACbKA, BBird, BD2412, BadWolf42, BadaBoom, Badger Brock, Balazer, Barts1a, Bass747, Baumi, Bdentremont, Beland, Belovedfreak, Ben-Zin, Benji Powell, Betacommand, Biblo, Bidabadi, Bigbluefish, Bigjimr, BilCat, Bilbobee, BillFlis, Binksternet, Birdhurst, Bjmullan, Black Widow, Blackanddarkness, Blackhawk1, Blackjackshellac, BlaiseFEgan, Blake Burba, Blake-, Blerg1, Blob5824, Bobblewik, Bolivian Unicyclist, Boundarylayer, Bovineone, Brennan626, Brianski, Brightgalrs, Brion VIBBER, Bristolhistorybuff, Britishairways, Britmax, Britoca, Brusegadi, Bryan Derksen, Bthebest, Burgundavia, Bzuk, C628, CDrecche, CP\M, CR85747, CU4ever, Cactus732, Calieber, CambridgeBayWeather, Camerong, Campdavid, Canterbury Tail, CapitalLetterBeginning, CaptRik, Captain Disdain, Captainbeefart, Carbonix, CardinalDan, CaribDigita, Caribpilot, Catgut, Charvex, Chase me ladies, I'm the Cavalry, Chaseincats, Chensiyuan, ChicXulub, Chirp Cricket, Choster, ChrisGualtieri, ChrisMD123, Chrisvandemore, Cjs2111, Clavecin, Cliffb, Clipper471, Cmd21700, Cmoney1998, Cointyro, Colmlinehan, CommonsDelinker, Compdude123, Conscious, Conversion script, Coolhawks88, Courcelles, Cowbert, Cpl Syx, Crispmuncher, Crohnie, Cvalda, Cygnus78, CygnusPius, D.c.camero, DJ Clayworth, DJ1AM, DMChatterton, DW, Dalillama, Dan 9790, Daniel Quinlan, DarkAvira, Darsys, Davehi1, David Corbett, David Latapie, David.Monniaux, Davidcannon, Davidjfarrar, Daytona2, Dbstommy, DeFacto, Deadlock, Deflective, Defsac, Delirium, Delldot, Deltabeignet, Deon, DerHexer, Derek Ross, DexDor, Dgrant, Dhaluza, Dhartung, Dhruvbird, Diderot, Digital1, DineshAdv, Dirk P Broer, Discomike, Discospinster, Disillusi, Dismas, Disphenoidal, Distantbody, DocBrown, DocendoDiscimus, Dockurt2k, Dodgethis, Dolovis, Dolphin51, DonPevsner, Dongledad, Donkdonk, Donreed, Doradus, DoubleBlue, Dowlingm, Dr. Blofeld, Dratman, Drbreznjev, Drbug, Driftwood87, Drmies, Drutt, Druzhnik, Dsr30, Dthomsen8, DuKot, Dual Freq, Dulciana, Dysprosia, ENeville, ESkog, Eanschuessler, Easterbradford, Eastpier, Easyup77, Ed g2s, EdC, EdChem, Edderso, Eddyy, Edmilne, Egil, Elmarko99, Emmaisthebest, Emoscopes, Emt147, Epbr123, ErgSlider, Eric Blatant, Eric Kvaalen, Ericd, Ericg, Ericoides, Ervin610, Esrever, Etimbo, Eurocopter, Eurosong, Everyguy, Everyking, Evilarry, Extransit, FAH Q 69, FClef, FT2, FaTaLERRORXX, Falcon8765, Fatboy1996, Favonian, Fawcett5, Felix116, Fences and windows, Ferrari5700, Fikri, Filanca, Firehawk1717, Fireswordfight, Floydian, Fnlayson, FocalPoint, Foxyshadis, Francs2000, Fraserspeirs, Freakmighty, Frerin, Frglee, Fripp gustafsson, Frymaster, FunkyFly, Fuzlyssa, GCarty, GDM, Gaius Cornelius, Galaxydog2000, Gaminrus, Ganeshema, Gareth E Kegg, Gbinal, Gcallard, Gene Nygaard, Gephart, Gimboid13, GliderMaven, Glitch82, Gmaxwell, Gobeirne, Goben10, GodGell, Gogo Dodo, GoingBatty, Gokaydince, Golbez, GoldDragon, Gparker, GraemeLeggett, Graft, GrahamN, Grebllaw123e, Green caterpillar, GregorB, Greyengine5, Griffinofwales, Grizzwald, Grondemar, Grover cleveland, Groyolo, Gsaup, Gsl, Gstekas, Guccigopackgo, GuiTaoist, Gunter, Gwen Gale, Gwernol, Gyrobo, Gzornenplatz, H1523702, HCT concorde engineering, HDCase, Hadal, Haham hanuka, Halsteadk, Hammersfan, Hans Dunkelberg, Harrison49, Harvestman, Hathaway36, Haya shiloh, Haymaker, Haymee, Hbdragon88, Hcp-07, He 2451, Head, Hede2000, Hektor, Heltzen, Herbie290, HiTek213, Homerjay, Hooperbloob, Htra0497, Hu, Huaiwei, Hubare, Hux, Hydeblake, Hydra Rider, IR Mongoose, Ian Dunster, IanWorthington, Icairns, Iconseditor, Igoldste, Imarquecm, Imjustmatthew, Immortal Synn, Impi, Incompetence, Inetpup, Infrogmation, InnocuousPseudonym, Iridescent, J.delanoy, JForget, JKL22, JRoman, JSira, Jack Upland, JackLumber, Jackwestmore, Jaganath, Jakers1304, Jam01, JamesBWatson, Jangles1, Jasepl, Jassy2010, Jatkins, Jawr256, Jay the Despicable, Jaywubba1887, Jcm, Jdmalouff, Jeff3000, Jeffrey Henning, Jemauvais, Jeronimo, JetBlast, JetLover, Jezhotwells, Jfitts, JiMidnite, JidGom, Jimjimeny, Jimmy Slade, Jimwilliams57, Jjacobsmeyer, JoeSperrazza, Joepane, John, John phillips235, John21Allen, Jojhutton, Jonboede, Jorbyma2007, Joseph Solis in Australia, Jpgordon, Jtdunlop, Julien.scavini, Junafani, Junky, JzG, Jérôme, K. Annoyomous, K1Bond007, KJBracey, KLM-Spotting-Service, Kaihsu, Kaldosh, Karl Dickman, Kazvorpal, Kb1koi, Kbthompson, Ken Gallager, Keraunoscopia, Khyth, Kimyu12, Kingboyk, Kingturtle, Kintetsubuffalo, Kirill Skorobogat, Kirk Hilliard, KittySaturn, Klaask, KnowledgeOfSelf, Kowloonese, Kozuch, Krellis, Kungfuadam, Kuru, Kwamikagami, Kyteto, L Kensington, LHOON, LWF, LachlanA, LanceBarber, Landroo, Lap dance with Dan, Lapinmies, LatinoFlava, Lazybum, LeaveSleaves, Lee M, LeedsKing, Lerdthenerd, Leslie Mateus, Leszek Jańczuk, Librarian2, Liftarn, Lightmouse, LilHelpa, Lincolnite, Lipton sale, Lisapalomeque, Locazreal, Lolgasm12345, LoopyB, Lorast, Lorax, Lotje, Lowellian, Lucy-marie, LukePearcy, Luna Santin, Lyricmac, Lyswim, M.nelson, M0RHI, M100, MBK004, MER-C, Mabdul, Maccoinnich, Macy, Madaboutrains, Magioladitis, Magnus Manske, Mailer diablo, Maksdo, Malcolmxl5, Malik54321, ManojMohamed, MapsMan, Marcika, Marek69, Mark Foskey, Mark83, MarkS, Marke14, Markonen, Marshall Stax, Martin villafuerte85, MartinVillafuerte85, Marvel52, Mathewignash, Matt Crypto, Mattfalcus, MatthewHaywood, Maury Markowitz, Mav, Maxí, Mazettas, Mboverload, McSly, Me125, Megiddo1013, Mel Etitis, Mellyrose, Memphisto, Mervyn, Mgw89, Mhkay, Mhking, MiShogun, Michael Devore, Michael Johnson, Michael Shields, Michael Zimmermann, MickMacNee, Mieciu K, Mike Rosoft, Mikus, MilborneOne, Miltonhowe, Mimono1997, Minesweeper, Mintguy, Miremare, Miterdale, Mjl0509, Mmx1, Modulatum, Mojei, Moncrief, Mono, Morganson691, Moriori, Morn, Morven, Motor, Mr Larrington, MrSpam, Mrceleb2007, Msa1701, Muad, Muckysock94, Mvjs, Myscrnnm, N328KF, N5iln, NarayanGa, Nasarius, Nathanjeffrey114, Nathans, NawlinWiki, Necessary Evil, NellieBly, NetRolller 3D, Neuro, Nibios, Nick Cooper, Nick45726, NicoBolso, Night Gyr, Nikai, Nikpapag, Nil Einne, Nono64, Nv8200p, Ohconfucius, Oliverdl, Olyus, OverlordQ, Oxyman42, Ozzieboy, Paaln, PaddyBriggs, Palamabron, Palizdesign, PapaBear1965, Paraphrased, Parnell88, Paronomasia, Patch'r5005, PatrickOfLondon, Patrickw421, Paul Richter, Paul.who, Pbmaise, Pcb21, Pdhroche, Peloneous, Pendo 4, Petri Krohn, Peyre, Phil75, Philg88, PhilipNewman, PierreAbbat, Pigsonthewing, Pinkbeast, Pittspilot, Pjc51, Planetscared, Pointillist, Poli08, PoliceChief, Populus, Postrach, Pratj, PrestonH, ProhibitOnions, Proprius, Provians, Prumpf, PsychoDave, Publicgirluk, Qleem, Qogir, QuadrivialMind, Quaeler, Qualit-E, QuiteUnusual, RG2, RTC, Rama, RandomStringOfCharacters, Rangoon11, RapidR, Rareeyes, Raryel, RaseaC, Raymondwinn, Rayward, Rccoms, Red, RedWolf, Redalert2fan, Redlock, Redthoreau, Redux, Reedy, Rehnn83, RekishiEJ, Rememberway, Rems 75, RenniePet, Rettetast, Rettig, Reubentg, Reywas92, Rhombus, Rich Farmbrough, Richc1977, Richi, Rickyrab, Ridow, Rillian, Ringwayobserver, Rjp uk, Rjwilmsi, Rlandmann, Rlcantwell, Rlove, Roadrunner, Rob690532, Rodw, Rollo, Ronhjones, Ronz, RoyBoy, Rphilipp, Rsnetto74, Rsrikanth05, Rumping, Russavia, RuthAS, S.prasad, SDC, SHCarter, SNIyer12, Sacxpert, Saint-Paddy, Salberti, Salfter, Sam Hocevar, SamH, SamatJain, Samslipknot, Samurai107, Sandover, Sandstein, Sbisolo, Scetoaux, ScienceApologist, Sciisfun, Scott Wilson, Sct72, Seanroxursox, Sebhaque, Seen46, Sekicho, Sepeario, Septegram, Setokaiba, Sfacets, Sfahey, Shadowjams, Shii, Shreditor, Shumtie, Sierra 1, SigmaEpsilon, SimonMayer, SimonP, Sir Nicholas de Mimsy-Porpington, Sjorford, Skamecrazy123, Skrelk, Skybon, Skydancer, Skywayman, Sl1982, Sladen, Slimey.limey, Smartse, Smurfwow, Snowmanradio, Snoyes, SoCalSuperEagle, Solipsist, Someth, Somethingoranother, Sophitus, Soumyasch, Sovxx, Sp33dyphil, Sparrowman980, Speight, Spellmaster, SpookyMulder, Ssbohio, Standardfact, StarryGrandma, Stebbiv, Steel, Steelpillow, Steinsky, Stemer1946, Stephen Hui, Stephenb, Stephenchou0722, Steschweiger, Stevo1000, Stewacide, Stilwebm, Str1977, Strayan, Stromcarlson, Super747, SuperJumbo, Superhacker2000, Superm401, SynergyStar, TGCP, Tango, Tannin, Tarquin, Tdadamemd, TeamAnand, Teapeat, Tec15, Tempshill, Teratornis, Tetracube, TexasAndroid, Texture, Tfpsly, Thadius856, The Bushranger, The Thing That Should Not Be, TheEgyptian, TheRanger, Theda, Thegrenade, Thermofan, Thevas, Thingg, Thinkpad, Thomasionus, Thumperward, Thunderbird2, Tiddly Tom, Tim!, TimS00, Timjarrett, Timo Kouwenhoven, Timrollpickering, Tinss, Tlesinski, Tobych, Tom, Tomwalden, TopGear44, Totensiebush, Tpbradbury, Trainspotter, Tranter, Trekphiler, Trev M, Trevor MacInnis, Tropdars, TrulyBlue, Trulystand700, Trusilver, Truthanado, Tumadoireacht, Tuzapicabit, Twobells, UB65, UKWiki, UkPaolo, Ukurko, Unara, Unregistered.coward, Unused0022, Urmomwithme1, UrsusArctosL71, Utgardsloki, V-100Commando, VK35, VampWillow, Vanished895703, Veej, Veesicle, Vgy7ujm, Viajero, Vicenarian, Vilcxjo, Vinceouca, Vishnava, VodkaJazz, Vssun, Vwozone, WDGraham, WLior, Wangi, Wapcaplet, Wavelength, Wcleak, Welshleprechaun, Wereon, WhisperToMe, WikHead, WikiDao, WikiuserNI, Will Beback Auto, Williamsze, Willking1979, Wispanow, Wizzy, Wmahan, Wolfkeeper, Woseph, Wrssrir, Wxidea, Wyss, X-mass, Xiaphias, Xmp, Xn4, Xtanstic, Xtzou, Xyzzyva, YSSYguy, Yesopenilno, Ylee, Yngvarr, YourEyesOnly, Ysangkok, Yvwv, Zantolak, Zero sharp, Zoicon5, Zoperd, Zzuuzz, Δάσκαλος Ιωσήφ, , 1712 anonymous edits

Concorde *Source*: http://en.wikipedia.org/w/index.php?title=Concorde *Contributors*: -Majestic-, 1995hoo, 1exec1, 200.191.188.xxx, 334a, 50pence, 621PWC, 790, 84user, 9913, A strolling player, A. Parrot, A380 Fan, ACY.Man, AD66, AEMoreira042281, AHands, AMCKen, ASV47, Abut, Acb58, Adimage, Adry1991, AdultSwim, Aeroeditin, Afowler, Ahoerstemeier, Ahunt, Aircorn, Airodyssey, Airplaneman, Ajkgordon, Ajray, Akradecki, Alai, Alansohn, Alaz, Albany NY, Aleenf1, Alex756, Alfio, Ali@gwc.org.uk, Alias777, Alsandro, Alvaro, Alvis, Amitkinger, Amplitude101, Amthomas0412, An2597, Andrejj, Andresihera, Android Mouse, Andy Dingley, Andy G, AnthonyUK, Antonio Lopez, AntonioMartin, Archzog, Aremith, Armeria, Arpingstone, AscendedAnathema, Ask123, Askari Mark, Aspensti, Asybaris01, Atlant, Auseplot, Autair, Avaya1, AxG, Axver, Ayla, BACbKA, BBird, BD2412, BadWolf42, BadaBoom, Badger Brock, Balazer, Barts1a, Bass747, Baumi, Bdentremont, Beland, Belovedfreak, Ben-Zin, Benji Powell, Betacommand, Biblo, Bidabadi, Bigbluefish, Bigjimr, BilCat, Bilbobee, BillFlis, Binksternet, Birdhurst, Bjmullan, Black Widow, Blackanddarkness, Blackhawk1, Blackjackshellac, BlaiseFEgan, Blake Burba, Blake-, Blerg1, Blob5824, Bobblewik, Bolivian Unicyclist, Boundarylayer, Bovineone, Brennan626, Brianski, Brightgalrs, Brion VIBBER, Bristolhistorybuff, Britishairways, Britmax, Britoca, Brusegadi, Bryan Derksen, Bthebest, Burgundavia, Bzuk, C628, CDrecche, CP\M, CR85747, CU4ever, Cactus732, Calieber, CambridgeBayWeather, Camerong, Campdavid, Canterbury Tail, CapitalLetterBeginning, CaptRik, Captain Disdain, Captainbeefart, Carbonix, CardinalDan, CaribDigita, Caribpilot, Catgut, Charvex, Chase me ladies, I'm the Cavalry, Chaseincats, Chensiyuan, ChicXulub, Chirp Cricket, Choster, ChrisGualtieri, ChrisMD123, Chrisvandemore, Cjs2111, Clavecin, Cliffb, Clipper471, Cmd21700, Cmoney1998, Cointyro, Colmlinehan, CommonsDelinker, Compdude123, Conscious, Conversion script, Coolhawks88, Courcelles, Cowbert, Cpl Syx, Crispmuncher, Crohnie, Cvalda, Cygnus78, CygnusPius, D.c.camero, DJ Clayworth, DJ1AM, DMChatterton, DW, Dalillama, Dan 9790, Daniel Quinlan, DarkAvira, Darsys, Davehi1, David Corbett, David Latapie, David.Monniaux, Davidcannon, Davidjfarrar, Daytona2, Dbstommy, DeFacto, Deadlock, Deflective, Defsac, Delirium, Delldot, Deltabeignet, Deon, DerHexer, Derek Ross, DexDor, Dgrant, Dhaluza, Dhartung, Dhruvbird, Diderot, Digital1, DineshAdv, Dirk P Broer, Discomike, Discospinster, Disillusi, Dismas, Disphenoidal, Distantbody, DocBrown, DocendoDiscimus, Dockurt2k, Dodgethis, Dolovis, Dolphin51, DonPevsner, Dongledad, Donkdonk, Donreed, Doradus, DoubleBlue, Dowlingm, Dr. Blofeld, Dratman, Drbreznjev, Drbug, Driftwood87, Drmies, Drutt, Druzhnik, Dsr30, Dthomsen8, DuKot, Dual Freq, Dulciana, Dysprosia, ENeville, ESkog, Eanschuessler, Easterbradford, Eastpier, Easyup77, Ed g2s, EdC, EdChem, Edderso, Eddyy, Edmilne, Egil, Elmarko99, Emmaisthebest, Emoscopes, Emt147, Epbr123, ErgSlider, Eric Blatant, Eric Kvaalen, Ericd, Ericg, Ericoides, Ervin610, Esrever, Etimbo, Eurocopter, Eurosong, Everyguy, Everyking, Evilarry, Extransit, FAH Q 69, FClef, FT2, FaTaLERRORXX, Falcon8765, Fatboy1996, Favonian, Fawcett5, Felix116, Fences and windows, Ferrari5700, Fikri, Filanca, Firehawk1717, Fireswordfight, Floydian, Fnlayson, FocalPoint, Foxyshadis, Francs2000, Fraserspeirs, Freakmighty, Frerin, Frglee,

Fripp gustafsson, Frymaster, FunkyFly, Fuzlyssa, GCarty, GDM, Gaius Cornelius, Galaxydog2000, Gaminrus, Ganeshema, Gareth E Kegg, Gbinal, Gcallard, Gene Nygaard, Gephart, Gimboid13, GliderMaven, Glitch82, Gmaxwell, Gobeirne, Goben10, GodGell, Gogo Dodo, GoingBatty, Gokaydince, Golbez, GoldDragon, Gparker, GraemeLeggett, Graft, GrahamN, Grebllaw123e, Green caterpillar, GregorB, Greyengine5, Griffinofwales, Grizzwald, Grondemar, Grover cleveland, Groyolo, Gsaup, Gsl, Gstekas, Guccigopackgo, GuiTaoist, Gunter, Gwen Gale, Gwernol, Gyrobo, Gzornenplatz, H1523702, HCT concorde engineering, HDCase, Hadal, Haham hanuka, Halsteadk, Hammersfan, Hans Dunkelberg, Harrison49, Harvestman, Hathaway36, Haya shiloh, Haymaker, Haymee, Hbdragon88, Hcp-07, He 2451, Head, Hede2000, Hektor, Heltzen, Herbie290, HiTek213, Homerjay, Hooperbloob, Htra0497, Hu, Huaiwei, Hubare, Hux, Hydeblake, Hydra Rider, IR Mongoose, Ian Dunster, IanWorthington, Icairns, Iconseditor, Igoldste, Imarquecm, Imjustmatthew, Immortal Synn, Impi, Incompetence, Inetpup, Infrogmation, InnocuousPseudonym, Iridescent, J.delanoy, JForget, JKL22, JRoman, JSira, Jack Upland, JackLumber, Jackwestmore, Jaganath, Jakers1304, Jam01, JamesBWatson, Jangles1, Jasepl, Jassy2010, Jatkins, Jawr256, Jay the Despicable, Jaywubba1887, Jcm, Jdmalouff, Jeff3000, Jeffrey Henning, Jemauvais, Jeronimo, JetBlast, JetLover, Jezhotwells, Jfitts, JiMidnite, JidGom, Jimjimeny, Jimmy Slade, Jimwilliams57, Jjacobsmeyer, JoeSperrazza, Joepane, John, John phillips235, John21Allen, Jojhutton, Jonboede, Jorbyma2007, Joseph Solis in Australia, Jpgordon, Jtdunlop, Julien.scavini, Junafani, Junky, JzG, Jérôme, K. Annoyomous, K1Bond007, KJBracey, KLM-Spotting-Service, Kaihsu, Kaldosh, Karl Dickman, Kazvorpal, Kb1koi, Kbthompson, Ken Gallager, Keraunoscopia, Khyth, Kimyu12, Kingboyk, Kingturtle, Kintetsubuffalo, Kirill Skorobogat, Kirk Hilliard, KittySaturn, Klaask, KnowledgeOfSelf, Kowloonese, Kozuch, Krellis, Kungfuadam, Kuru, Kwamikagami, Kyteto, L Kensington, LHOON, LWF, LachlanA, LanceBarber, Landroo, Lap dance with Dan, Lapinmies, LatinoFlava, Lazybum, LeaveSleaves, Lee M, LeedsKing, Lerdthenerd, Leslie Mateus, Leszek Jańczuk, Librarian2, Liftarn, Lightmouse, LilHelpa, Lincolnite, Lipton sale, Lisapalomeque, Locazreal, Lolgasm12345, LoopyB, Lorast, Lorax, Lotje, Lowellian, Lucy-marie, LukePearcy, Luna Santin, Lyricmac, Lyswim, M.nelson, M0RHI, M100, MBK004, MER-C, Mabdul, Maccoinnich, Macy, Madaboutrains, Magioladitis, Magnus Manske, Mailer diablo, Maksdo, Malcolmxl5, Malik54321, ManojMohamed, MapsMan, Marcika, Marek69, Mark Foskey, Mark83, MarkS, Marke14, Markonen, Marshall Stax, Martin villafuerte85, MartinVillafuerte85, Marvel52, Mathewignash, Matt Crypto, Mattfalcus, MatthewHaywood, Maury Markowitz, Mav, Maxí, Mazettas, Mboverload, McSly, Me125, Megiddo1013, Mel Etitis, Mellyrose, Memphisto, Mervyn, Mgw89, Mhkay, Mhking, MiShogun, Michael Devore, Michael Johnson, Michael Shields, Michael Zimmermann, MickMacNee, Mieciu K, Mike Rosoft, Mikus, MilborneOne, Miltonhowe, Mimono1997, Minesweeper, Mintguy, Miremare, Miterdale, Mjl0509, Mmx1, Modulatum, Mojei, Moncrief, Mono, Morganson691, Moriori, Morn, Morven, Motor, Mr Larrington, MrSpam, Mrceleb2007, Msa1701, Muad, Muckysock94, Mvjs, Myscrnnm, N328KF, N5iln, NarayanGa, Nasarius, Nathanjeffrey114, Nathans, NawlinWiki, Necessary Evil, NellieBly, NetRolller 3D, Neuro, Nibios, Nick Cooper, Nick45726, NicoBolso, Night Gyr, Nikai, Nikpapag, Nil Einne, Nono64, Nv8200p, Ohconfucius, Oliverdl, Olyus, OverlordQ, Oxyman42, Ozzieboy, Paaln, PaddyBriggs, Palamabron, Palizdesign, PapaBear1965, Paraphrased, Parnell88, Paronomasia, Patch'r5005, PatrickOfLondon, Patrickw421, Paul Richter, Paul.who, Pbmaise, Pcb21, Pdhroche, Peloneous, Pendo 4, Petri Krohn, Peyre, Phil75, Philg88, PhilipNewman, PierreAbbat, Pigsonthewing, Pinkbeast, Pittspilot, Pjc51, Planetscared, Pointillist, Poli08, PoliceChief, Populus, Postrach, Pratj, PrestonH, ProhibitOnions, Proprius, Provians, Prumpf, PsychoDave, Publicgirluk, Qleem, Qogir, QuadrivialMind, Quaeler, Qualit-E, QuiteUnusual, RG2, RTC, Rama, RandomStringOfCharacters, Rangoon11, RapidR, Rareeyes, Raryel, RaseaC, Raymondwinn, Rayward, Rccoms, Red, RedWolf, Redalert2fan, Redlock, Redthoreau, Redux, Reedy, Rehnn83, RekishiEJ, Rememberway, Rems 75, RenniePet, Rettetast, Rettig, Reubentg, Reywas92, Rhombus, Rich Farmbrough, Richc1977, Richi, Rickyrab, Ridow, Rillian, Ringwayobserver, Rjp uk, Rjwilmsi, Rlandmann, Rlcantwell, Rlove, Roadrunner, Rob690532, Rodw, Rollo, Ronhjones, Ronz, RoyBoy, Rphilipp, Rsnetto74, Rsrikanth05, Rumping, Russavia, RuthAS, S.prasad, SDC, SHCarter, SNIyer12, Sacxpert, Saint-Paddy, Salberti, Salfter, Sam Hocevar, SamH, SamatJain, Samslipknot, Samurai107, Sandover, Sandstein, Sbisolo, Scetoaux, ScienceApologist, Sciisfun, Scott Wilson, Sct72, Seanroxursox, Sebhaque, Seen46, Sekicho, Sepeario, Septegram, Setokaiba, Sfacets, Sfahey, Shadowjams, Shii, Shreditor, Shumtie, Sierra 1, SigmaEpsilon, SimonMayer, SimonP, Sir Nicholas de Mimsy-Porpington, Sjorford, Skamecrazy123, Skrelk, Skybon, Skydancer, Skywayman, Sl1982, Sladen, Slimey.limey, Smartse, Smurfwow, Snowmanradio, Snoyes, SoCalSuperEagle, Solipsist, Someth, Somethingoranother, Sophitus, Soumyasch, Sovxx, Sp33dyphil, Sparrowman980, Speight, Spellmaster, SpookyMulder, Ssbohio, Standardfact, StarryGrandma, Stebbiv, Steel, Steelpillow, Steinsky, Stemer1946, Stephen Hui, Stephenb, Stephenchou0722, Steschweiger, Stevo1000, Stewacide, Stilwebm, Str1977, Strayan, Stromcarlson, Super747, SuperJumbo, Superhacker2000, Superm401, SynergyStar, TGCP, Tango, Tannin, Tarquin, Tdadamemd, TeamAnand, Teapeat, Tec15, Tempshill, Teratornis, Tetracube, TexasAndroid, Texture, Tfpsly, Thadius856, The Bushranger, The Thing That Should Not Be, TheEgyptian, TheRanger, Theda, Thegrenade, Thermofan, Thevas, Thingg, Thinkpad, Thomasionus, Thumperward, Thunderbird2, Tiddly Tom, Tim!, TimS00, Timjarrett, Timo Kouwenhoven, Timrollpickering, Tinss, Tlesinski, Tobych, Tom, Tomwalden, TopGear44, Totensiebush, Tpbradbury, Trainspotter, Tranter, Trekphiler, Trev M, Trevor MacInnis, Tropdars, TrulyBlue, Trulystand700, Trusilver, Truthanado, Tumadoireacht, Tuzapicabit, Twobells, UB65, UKWiki, UkPaolo, Ukurko, Unara, Unregistered.coward, Unused0022, Urmomwithme1, UrsusArctosL71, Utgardsloki, V-100Commando, VK35, VampWillow, Vanished895703, Veej, Veesicle, Vgy7ujm, Viajero, Vicenarian, Vilcxjo, Vinceouca, Vishnava, VodkaJazz, Vssun, Vwozone, WDGraham, WLior, Wangi, Wapcaplet, Wavelength, Wcleak, Welshleprechaun, Wereon, WhisperToMe, WikHead, WikiDao, WikiuserNI, Will Beback Auto, Williamsze, Willking1979, Wispanow, Wizzy, Wmahan, Wolfkeeper, Woseph, Wrssrir, Wxidea, Wyss, X-mass, Xiaphias, Xmp, Xn4, Xtanstic, Xtzou, Xyzzyva, YSSYguy, Yesopenilno, Ylee, Yngvarr, YourEyesOnly, Ysangkok, Yvwv, Zantolak, Zero sharp, Zoicon5, Zoperd, Zzuuzz, Δάσκαλος Ιωσήφ, , 1712 anonymous edits

BT_Group *Source*: http://en.wikipedia.org/w/index.php?title=BT_Group *Contributors*: 146.101.149.xxx, 159753, 195.99.244.xxx, 1exec1, 220 of Borg, 2toy mora, 5 albert square, 76John76, 78judgered, AJR, ALR, AMPJA, AMe, Abu-Fool Danyal ibn Amir al-Makhiri, Abuttle, Acroterion, Aeons, AgarwalSumeet, Albino.pure, Alika12, Allyl90, Alphonse2, Amy Crescenzo, Andrews Palop, Animum, Apers0n, Apotheosis247, Arletty75, Arthur Frayn, Asyndeton, AussieLegend, Autolycus, Avebury, AxG, Backchannel, Barek, Basys, Bbhart, BenShade, Bencherlite, Benson85, Bigjimr, Bigrich, Biker Biker, BlaiseFEgan, Bobblewik, Bobrayner, Bongomatic, Bonusballs, Bornintheguz, BradM, Brandsen, Breadandcheese, Brett Dunbar, Brian McNeil, Bryan Derksen, Btglobalservices, Buckstars, Bungofpot, Burda001, Burto88, CGorman, CR7, Cameron8276, Canderra, Carina22, Catapult, Cdanger, Cefn, Chris the speller, ChrisGualtieri, Chrisieboy, Chriswiki, Cloudbound, Cmdrjameson, Control-alt-delete, Conversion script, Coolhawks88, Cosmicpop, Cyanoa Crylate, Cybercobra, D 2 da a, DMac1919, Damo2929, Daniel, Danielgwj, Danwareuk123, Darkieboy236, Designcouncil, Df747jet, Dispenser, Dman727, Doc glasgow, DocendoDiscimus, DominicConnor, Dormskirk, DoruMarx, Dpaajones, EGC, EditorInTheRye, Editorial1, Edwy, Emoscopes, Ephebi, Etenebris, F20ljc, Fairlyoddparents1234, Falcon8765, FlecherA, Fran B. E., FredOrAlive, Frietjes, Fuzheado, Fuzzie, GameKeeper, Geosynchronous, GermanUser2045, Giggedypigedy, Gimboid13, Gobonobo, Godlesswanderer, Graham87, Greenshed, Gryffindor, Gurch, Guruparan18, H1523702, H2g2bob, Hairy Dude, Happysailor, Harryzilber, Hielant, Hu12, I need a name, Ian3055, Imagi-King, Imakeaccountssofunnynamesendupintheedithistory, Immunize, Iridescent, IronGargoyle, J.P.Lon, JD554, JDCMAN, Jaffacakesrgreat, Jamcib, James500, Jamesa1983, Jamesinderbyshire, Jamirocrazy, Jasonfward, Jayjg, Jdforrester, Jean.julius, Jeff Song, Jeni, Jennavecia, Jerryseinfeld, JetBlast, JimjReilly, Jjdejong, Jjp43, John, John a s, John of Reading, Jojalozzo, JonathanDP81, Jonralton, Joolz, Josh Parris, Josh477, Joshii, Joshnicholson, Jrockley, JulianHensey, Kaihsu, Kazoozoo, Keith D, Khendon, Kiand, Kileg, Kilo-Lima, Kingsixdx, Kjkolb, Kkm010, KnowledgeOfSelf, Koavf, Kozuch, KuroiShiroi, Kwamikagami, L353a1, Lawsonrob1, LeaveSleaves, LeedsKing, Lightmouse, LittleWink, Londoner 2008, Lotje, Lumos3, MER-C, Mark126, Mark83, Mark91, Mark999, Marknew, Matt Crypto, Mattbr, Matthew Platts, Mauls, Max Naylor, Mdd, MementoVivere, Mervyn, Mets501, Mhaswell, Mhockey, Mhoulden, Michaelpuk1, MichellePollock1245, Mirv, Mjwalshe, Moe Epsilon, Moonriddengirl, MrWikiTroll, Mwongozi, Mxn, Nescio vos, Neurolysis, Neutrality, Niceguyedc, Nigelj, Ninetyone, Nixdorf, No1lakersfan, NoOne898, Nonomy, NotMuchToSay, Ohconfucius, Oliver Lineham, Olivier, Omicronpersei8, OneTwoFour, Oxymoron83, Paoloattivissimo, Pcpcpc, Pearcer, Pearle, Percy raja, Peridon, Phe, PhiJ, Phil Holmes, PhilipO, Piano non troppo, Pinkadelica, Polainm, Pontificalibus, ProhibitOnions, Psantora, Quadduc, QubitOtaku, R'n'B, RFBailey, RM21, Rafert, Randomaniac, Rangoon11, Rapido, Rbrwr, Rdsmith4, Redvers, Reflex Reaction, Requestion, RetiredUser124642196, Rich Farmbrough, Rjwilmsi, Rmz, Robert Brook, Rose Booth, Russpear, Rusticsmile, Rwsansom, RyanGerbil10, SD5, SMC89, Saikokira, Scott Bywater, SeanMack, Sebcachia, Secretlondon, Seth Whales, Shawnc, Sherool, Sianwj, Sibruk, SimonTrew, Simple Bob, Sir Stanley, Sjc, Sloman, Sm07c, Smjg, Solipsist, Sonance, Speed2006, Speer320, Sprocket, Starflyer, StaticGull, Stephenb, Stevenlovespigs130793, Stoff, Stokesd3, Struway2, Stwalkerster, Sully, Svgalbertian, Tabletop, TaerkastUA, Tagishsimon, Tascha96, Technik, Techtri, The Anome, The Dark Arrg, The Nut, The sock that should not be, The wub, TheEnlightened, Thedreamdied, Thegn, Thehelpfulone, ThinkingTwice, ThomFewkes, Thorpe, Tim!, Tolien, Tom Morris, Tombomp, Tony Corsini, Tony1, Trademark ted, Trident13, Triwbe, Tubezone, Tvarma, UKintheUS, UkPaolo, Varlaam, Velella, Vivenot, WDGraham, Waspboy, Wellbelove, Wereon, WhisperToMe, Who141, Wiki alf, William Graham, Wimt, Woohookitty, Yurik, Zzuuzz, 601 anonymous edits

Image Sources, Licenses and Contributors

Image:1stViscountHanworth.jpg *Source*: http://en.wikipedia.org/w/index.php?title=File:1stViscountHanworth.jpg *License*: unknown *Contributors*: Vandyk Ltd.

File:Glastonbury 2008 126.JPG *Source*: http://en.wikipedia.org/w/index.php?title=File:Glastonbury_2008_126.JPG *License*: unknown *Contributors*: User:Ronhjones

File:Tony Benn 2005.jpg *Source*: http://en.wikipedia.org/w/index.php?title=File:Tony_Benn_2005.jpg *License*: unknown *Contributors*: User:JK the Unwise

File:Tony Benn Giles Fraser Levellers Day Burford 20080517.jpg *Source*: http://en.wikipedia.org/w/index.php?title=File:Tony_Benn_Giles_Fraser_Levellers_Day_Burford_20080517.jpg *License*: unknown *Contributors*: User:Kaihsu

File:Dartford Living September 2009 Front Cover.jpg *Source*: http://en.wikipedia.org/w/index.php?title=File:Dartford_Living_September_2009_Front_Cover.jpg *License*: unknown *Contributors*: User:Dartfordliving

File:Tony Benn, Tolpuddle 2012.JPG *Source*: http://en.wikipedia.org/w/index.php?title=File:Tony_Benn,_Tolpuddle_2012.JPG *License*: unknown *Contributors*: User:Rwendland

File:AbpCharlesThomasLongley.jpg *Source*: http://en.wikipedia.org/w/index.php?title=File:AbpCharlesThomasLongley.jpg *License*: unknown *Contributors*: Contemporary portrait

File:Charles Thomas Longley.jpg *Source*: http://en.wikipedia.org/w/index.php?title=File:Charles_Thomas_Longley.jpg *License*: unknown *Contributors*: User:Dcoetzee

File:Charles Thomas Longley by George Richmond.jpg *Source*: http://en.wikipedia.org/w/index.php?title=File:Charles_Thomas_Longley_by_George_Richmond.jpg *License*: unknown *Contributors*: MarmadukePercy

Image:1st Viscount Cobham.png *Source*: http://en.wikipedia.org/w/index.php?title=File:1st_Viscount_Cobham.png *License*: unknown *Contributors*: Original uploader was Craigy144 at en.wikipedia

Image:Stowe Armorial.jpg *Source*: http://en.wikipedia.org/w/index.php?title=File:Stowe_Armorial.jpg *License*: unknown *Contributors*: P. Sonard

File:Liberal Party logo (pre1988).png *Source*: http://en.wikipedia.org/w/index.php?title=File:Liberal_Party_logo_(pre1988).png *License*: unknown *Contributors*: MGA73

File:Leeds Town Hall, General Election results.jpg *Source*: http://en.wikipedia.org/w/index.php?title=File:Leeds_Town_Hall,_General_Election_results.jpg *License*: unknown *Contributors*: Unknown (date stipulates piece makes terms of 70 year rule)

Image:palmerston.jpg *Source*: http://en.wikipedia.org/w/index.php?title=File:Palmerston.jpg *License*: unknown *Contributors*: Anton Kazmyarchuk, Conscious, Dunwich, Samulili

Image:Acgladstone2.jpg *Source*: http://en.wikipedia.org/w/index.php?title=File:Acgladstone2.jpg *License*: unknown *Contributors*: Hsarrazin, Kilom691, Nicke L, Shakko, , 1 anonymous edits

Image:Asquith Q 42036.jpg *Source*: http://en.wikipedia.org/w/index.php?title=File:Asquith_Q_42036.jpg *License*: unknown *Contributors*: British Government

Image:David Lloyd George.jpg *Source*: http://en.wikipedia.org/w/index.php?title=File:David_Lloyd_George.jpg *License*: unknown *Contributors*: Harris & Ewing

Image:UK popular vote.svg *Source*: http://en.wikipedia.org/w/index.php?title=File:UK_popular_vote.svg *License*: unknown *Contributors*: User:Kanguole

Image:Gws samuel 01.jpg *Source*: http://en.wikipedia.org/w/index.php?title=File:Gws_samuel_01.jpg *License*: unknown *Contributors*: Senemmar

File:SDP Logo.png *Source*: http://en.wikipedia.org/w/index.php?title=File:SDP_Logo.png *License*: unknown *Contributors*: Good Olfactory, Yorkshirian

File:David Owen.jpg *Source*: http://en.wikipedia.org/w/index.php?title=File:David_Owen.jpg *License*: unknown *Contributors*: London School of Economics Library

File:RobertMacLennan1987 cropped.jpg *Source*: http://en.wikipedia.org/w/index.php?title=File:RobertMacLennan1987_cropped.jpg *License*: unknown *Contributors*: User:BartBassist, User:Rodhullandemu

File:British Airways Concorde G-BOAC 03.jpg *Source*: http://en.wikipedia.org/w/index.php?title=File:British_Airways_Concorde_G-BOAC_03.jpg *License*: unknown *Contributors*: Eduard Marmet

File:Concorde on Bristol.jpg *Source*: http://en.wikipedia.org/w/index.php?title=File:Concorde_on_Bristol.jpg *License*: unknown *Contributors*: User:Arpingstone

File:British Concorde.jpg *Source*: http://en.wikipedia.org/w/index.php?title=File:British_Concorde.jpg *License*: unknown *Contributors*: User:Plismo

File:Concorde first visit Heathrow Fitzgerald.jpg *Source*: http://en.wikipedia.org/w/index.php?title=File:Concorde_first_visit_Heathrow_Fitzgerald.jpg *License*: unknown *Contributors*: Steve Fitzgerald

File:ConcordeCockpitSinsheim.jpg *Source*: http://en.wikipedia.org/w/index.php?title=File:ConcordeCockpitSinsheim.jpg *License*: unknown *Contributors*: Ariadacapo, Dmitry G, Gomera-b, Helicop, KTo288, Marcelloo, ONC, PeterWD, Pullus In Fabula, Soeren Gasch, 1 anonymous edits

Image:2001concordewingduxfordJM.jpg *Source*: http://en.wikipedia.org/w/index.php?title=File:2001concordewingduxfordJM.jpg *License*: unknown *Contributors*: User:Guinnog

Image:Concordeintake.gif *Source*: http://en.wikipedia.org/w/index.php?title=File:Concordeintake.gif *License*: unknown *Contributors*: Original uploader was Burbank at en.wikipedia.

Image:Concorde Ramp.jpg *Source*: http://en.wikipedia.org/w/index.php?title=File:Concorde_Ramp.jpg *License*: unknown *Contributors*: Original uploader was Dockurt2k at en.wikipedia

File:Concorde fuel trim.svg *Source*: http://en.wikipedia.org/w/index.php?title=File:Concorde_fuel_trim.svg *License*: unknown *Contributors*: User:Steal88

Image:ConcordeFuselageSinsheim.jpg *Source*: http://en.wikipedia.org/w/index.php?title=File:ConcordeFuselageSinsheim.jpg *License*: unknown *Contributors*: User:Helicop

Image:Concorde interior2.jpg *Source*: http://en.wikipedia.org/w/index.php?title=File:Concorde_interior2.jpg *License*: unknown *Contributors*: This image was taken by Martin J.Galloway. Original uploader was Dotonegroup at en.wikipedia

File:Concorde at Baginton - geograph.org.uk - 156846.jpg *Source*: http://en.wikipedia.org/w/index.php?title=File:Concorde_at_Baginton_-_geograph.org.uk_-_156846.jpg *License*: unknown *Contributors*: Ardfern, Ariadacapo, Man vyi, Michael Barera, Mtaylor848, PeterWD

Image:Train d'atterrissage Concorde Musee du Bourget P1020039.JPG *Source*: http://en.wikipedia.org/w/index.php?title=File:Train_d'atterrissage_Concorde_Musee_du_Bourget_P1020039.JPG *License*: unknown *Contributors*: User:Pline

Image:Concorde tail gear.JPG *Source*: http://en.wikipedia.org/w/index.php?title=File:Concorde_tail_gear.JPG *License*: unknown *Contributors*: User:Mgw89

File:Concorde landing Farnborough Fitzgerald.jpg *Source*: http://en.wikipedia.org/w/index.php?title=File:Concorde_landing_Farnborough_Fitzgerald.jpg *License*: unknown *Contributors*: Steve Fitzgerald

File:British Airways Concorde official handover ceremony Fitzgerald.jpg *Source*: http://en.wikipedia.org/w/index.php?title=File:British_Airways_Concorde_official_handover_ceremony_Fitzgerald.jpg *License*: unknown *Contributors*: Steve Fitzgerald

File:Singapore Airlines Concorde Fitzgerald-1.jpg *Source*: http://en.wikipedia.org/w/index.php?title=File:Singapore_Airlines_Concorde_Fitzgerald-1.jpg *License*: unknown *Contributors*: Steve Fitzgerald

File:Concorde 1977.jpg *Source*: http://en.wikipedia.org/w/index.php?title=File:Concorde_1977.jpg *License*: unknown *Contributors*: PH3 Caffaro

File:Concorde at JFK.jpg *Source*: http://en.wikipedia.org/w/index.php?title=File:Concorde_at_JFK.jpg *License*: unknown *Contributors*: Hunter Desportes

File:concorde g-boab in storage arp.jpg *Source*: http://en.wikipedia.org/w/index.php?title=File:Concorde_g-boab_in_storage_arp.jpg *License*: unknown *Contributors*: Apalsola, Ardfern, Arpingstone, Edward, Gomera-b, KTo288, Kyteto, Mtaylor848, Oddbodz, Ronaldino

File:ConcordeCDG.jpg *Source*: http://en.wikipedia.org/w/index.php?title=File:ConcordeCDG.jpg *License*: unknown *Contributors*: Original uploader was Eurocopter at en.wikipedia. Later version(s) were uploaded by Arpingstone at en.wikipedia.

File:Conc01.jpg *Source*: http://en.wikipedia.org/w/index.php?title=File:Conc01.jpg *License*: unknown *Contributors*: Original uploader was PaddyBriggs at en.wikipedia

File:Concorde at Manchester Viewing Park.jpg *Source*: http://en.wikipedia.org/w/index.php?title=File:Concorde_at_Manchester_Viewing_Park.jpg *License*: unknown *Contributors*: Kurt Trew from Truro, United Kingdom

Image:ConcordeBG.jpg *Source*: http://en.wikipedia.org/w/index.php?title=File:ConcordeBG.jpg *License*: unknown *Contributors*: User:XOX

Image:Queen Elizabeth II and Prince Philip disembark from a British Airways Concorde.jpg *Source*: http://en.wikipedia.org/w/index.php?title=File:Queen_Elizabeth_II_and_Prince_Philip_disembark_from_a_British_Airways_Concorde.jpg *License*: unknown *Contributors*: SRA JERRY WILSON

File:Concordev1.0.png *Source*: http://en.wikipedia.org/w/index.php?title=File:Concordev1.0.png *License*: unknown *Contributors*: User:Julien.scavini

File:Concorde G-BOAC.png *Source*: http://en.wikipedia.org/w/index.php?title=File:Concorde_G-BOAC.png *License*: unknown *Contributors*: User:Emoscopes

Image:BT logo.svg *Source*: http://en.wikipedia.org/w/index.php?title=File:BT_logo.svg *License*: unknown *Contributors*: Avicennasis, AxG

File:BT Tower-1.jpg *Source*: http://en.wikipedia.org/w/index.php?title=File:BT_Tower-1.jpg *License*: unknown *Contributors*: Ardfern, Dcastor, Mtaylor848

File:BT Centre 3195 stitched.jpg *Source*: http://en.wikipedia.org/w/index.php?title=File:BT_Centre_3195_stitched.jpg *License*: unknown *Contributors*: User:Gryffindor

File:BT Openreach van.jpg *Source*: http://en.wikipedia.org/w/index.php?title=File:BT_Openreach_van.jpg *License*: unknown *Contributors*: User:Kaihsu

File:BT Adastral Park, Ipswich, UK - geograph.org.uk - 39100.jpg *Source*: http://en.wikipedia.org/w/index.php?title=File:BT_Adastral_Park,_Ipswich,_UK_-_geograph.org.uk_-_39100.jpg *License*: unknown *Contributors*: Nigel Freeman

Image:British Telecom 1981.svg *Source*: http://en.wikipedia.org/w/index.php?title=File:British_Telecom_1981.svg *License*: unknown *Contributors*: BT

Image:BT old logo.svg *Source*: http://en.wikipedia.org/w/index.php?title=File:BT_old_logo.svg *License*: unknown *Contributors*: Avicennasis, Melesse, Neurolysis, 2 anonymous edits

MIX
Papier aus verantwortungsvollen Quellen
Paper from responsible sources
FSC® C105338

Printed by Books on Demand GmbH, Norderstedt / Germany